普通高等教育农业农村部"十三五"规划教材
全国高等农林院校"十三五"规划教材

农业科技英语综合教程

第 二 版

孟 静 岳本杰 姜丽丽 主编

中国农业出版社
北 京

图书在版编目（CIP）数据

农业科技英语综合教程／孟静，岳本杰，姜丽丽主编．—2版．—北京：中国农业出版社，2022.2
普通高等教育农业农村部"十三五"规划教材　全国高等农林院校"十三五"规划教材
ISBN 978-7-109-29192-8

Ⅰ.①农… Ⅱ.①孟…②岳…③姜… Ⅲ.①农业科学－英语－高等学校－教材　Ⅳ.①S

中国版本图书馆CIP数据核字（2022）第040145号

农业科技英语综合教程
NONGYE KEJI YINGYU ZONGHE JIAOCHENG

中国农业出版社出版
地址：北京市朝阳区麦子店街18号楼
邮编：100125
策划编辑：马颉晨　龙永志　　责任编辑：马颉晨
版式设计：杜　然　　责任校对：刘丽香
印刷：北京通州皇家印刷厂
版次：2015年8月第1版　　2022年2月第2版
印次：2022年2月第2版北京第1次印刷
发行：新华书店北京发行所
开本：720mm×960mm　1/16
印张：11.5
字数：206千字
定价：27.00元

版权所有·侵权必究
凡购买本社图书，如有印装质量问题，我社负责调换。
服务电话：010-59195115　010-59194918

第二版编写人员

主　编　孟　静　岳本杰　姜丽丽
副主编　范媛媛　丁志锐　李秀杰
参　编　刘淑波　郭继鹏　董丽丽
　　　　　陈俊杰　杨　蕾　杜　曼
　　　　　贺清林　魏　晴　张龙珠

第一版编写人员

主　编　宋银秋　王月露　张天飞
副主编　刘　伟　王　彬　丁志锐
参　编　（按姓名笔画排序）
　　　　张玉东　范媛媛　姜丽丽

第二版前言

2020年7月全国研究生教育大会的召开，揭开了我国研究生教育的新篇章。大会结束之后，教育部、国家发展和改革委员会、财政部三部委在9月联合发布《关于加快新时代研究生教育改革发展的意见》，明确新时期我国研究生教育发展的路径、举措和目标等内容。作为研究生教育的一部分，研究生公共英语教学在培养学生跨文化沟通能力、国际学术合作与发表能力等方面发挥着重要作用。

目前研究生英语教学仍然偏重公共英语教学，教学目标不能适应时代的需求，教学内容与本科教学重复，教学效果不理想。在社会对研究生专业型外语知识及技能提出更高要求的背景下，研究生英语教学面临着新的挑战。为了适应新的发展形势，全面提高农业院校研究生英语的综合运用能力，我们对《农业科技英语综合教程》进行了修订。

教材特点

一、选材合理。本教材选材主要来自国际科研、学术机构，国际组织官网和国内现有教材，其目的是让农业及其相关专业学习者、从业者能够全面了解农业领域的背景知识和科学研究、学术实践活动，掌握国际学术现状、发展动态等宏观性农业信息。

二、内容充实。本教材共八个单元。每个单元分为Text A、Text B、科技文体翻译技巧、科技英语摘要写作和摘要译写示例五个部分。本教材对传统的公共外语语言进行瘦身，即向"大农业类公共外语"转型，并致力于衔接各学科类别中的"专业英语"。本教材在内容上涵盖了大农业背景的语言文化知识，是带有一定农业专业性的科技型或学术类的英语教材。

三、讲练结合。每个单元 Text A 后除单词外，还列出了相关专业词汇，Text B 后则是阅读理解，之后包含翻译和摘要修改等。翻译和摘要部分囊括了大部分科技文体经常使用的技巧和方法。摘要译写示例部分充分展示了如何正确修改摘要，并按照各单元相关专业编排，针对性更强。

教材定位

本教材设计的研习层次是大学英语四级以上水平的语言学习者，属于英语类阅读教程，适合于高等农业院校的本科生、硕士生和农业科研领域从业人员教学和研读。本教材基本涵盖了大农业一级学科的各个方向，每个学科的从业者和学习者都能找到可以借鉴的工具性网站和资料来源，相关学科研究者也可以在更广泛领域进行跨学科的研究学习。从英语语言文化学习的角度看，本教材中的英语文化更倾向于农业学科文化，对学习者、管理者、研究者和从业者皆有涉及，应该说是一部大农业类的英语语言文化学习的小型工具书。

由于本教材涉及大农业类学科面较广，学科间的语言表达具有一定差异性，在编写过程中专业知识的疏漏之处在所难免，敬请各学科专家和广大读者给予指正。

编 者

2021 年 10 月

第一版前言

自 2011 年 10 月全国农科院校研究生英语教学研究会召开以来，研究生英语教学改革的呼声日渐强烈。目前研究生英语教学大多沿袭多年以来的做法：偏重公共英语教学，教学目标不适应时代的需求，教学内容与本科教学重复，教学效果不理想。在社会对研究生专业型外语知识及技能提出更高要求的背景下，研究生英语教学面临着新的挑战。

为了适应新的发展形势，培养研究生用英语读写学术文献和技术资料的技能，全面提高农业院校研究生英语的综合运用能力，我们编写了这本《农业科技英语综合教程》。

一、教材特点

1. 选材合理。本教材选材主要来自国际科研、学术机构和国际组织官网及国内现有教材，其目的是让农业学习者、从业者能够全面了解农业领域的背景知识和科学研究、学术实践活动，掌握国际学术现状、发展动态等宏观性农业信息。

2. 内容充实。本教材共八个单元。每个单元分为 Text A、Text B、科技文体翻译技巧、科技英语摘要写作和摘要译写示例五个部分。本教材将传统的公共外语语言瘦身，向"大农业类公共外语"转型，并致力于衔接各学科类别中的"专业英语"。本教材在内容上是涵盖了大农业背景的语言文化知识，并带有一定农业专业性的科技型或学术类的英语教材。

3. 讲练结合。每个单元 Text A 后除单词外，列出了相关专业词汇，Text B 后是阅读理解，之后包含翻译和摘要修改等。翻译和摘要部分囊括了大部分科技文体经常使用的技巧和方法。摘要译写示例部分充分展示了如何正确修改摘要，并按照各单元相关专业编排，针对性更强。

二、教材定位

本教材设计的研习层次是大学英语四级以上水平的语言学习者，属于英语类阅读教程，适合于高等农业院校的本科生、硕士生和农业科研领域从业人员教学和研读。本教材基本涵盖了大农业一级学科的各个方向，每个学科的从业者和学习者都能找到可以借鉴的工具性网站和资料来源，相关学科研究者也可以在更广泛领域进行跨学科的研究学习。从英语语言文化学习角度看，本教材所涉及的英语文化则完全倾向于农业学科文化，对学习者、管理者、研究者和从业者皆有涉及，应该说是一部大农业类的英语语言文化学习的小型工具书。

由于本教材涉及大农业类学科面较广，学科间的语言表达具有一定差异性，在编写过程中专业知识的疏漏之处在所难免，敬请各学科专家和广大读者批评指正。

编 者

2015 年 1 月

目　　录

第二版前言
第一版前言

Unit 1　Agronomy ·· 1
　Text A ·· 1
　Text B ·· 8
　科技文体翻译技巧（一）　翻译的增减 ·· 13
　科技英语摘要写作（一）　信息型摘要 ·· 16
　摘要译写示例 ··· 18

Unit 2　Traditional Chinese Medicine ·· 22
　Text A ·· 22
　Text B ·· 31
　科技文体翻译技巧（二）　翻译的转换 ·· 35
　科技英语摘要写作（二）　指示型摘要 ·· 39
　摘要译写示例 ··· 39

Unit 3　Education ·· 43
　Text A ·· 43
　Text B ·· 51
　科技文体翻译技巧（三）　从句的译法 ·· 56
　科技英语摘要写作（三）　指示-信息型摘要 ·· 62
　摘要译写示例 ··· 63

Unit 4　Environment ·· 67
　Text A ·· 67
　Text B ·· 74
　科技文体翻译技巧（四）　特殊词的译法 ·· 80

科技英语摘要写作（四）　摘要的主题句 …………………………… 83
摘要译写示例 ………………………………………………………… 84

Unit 5　Sociology …………………………………………………… 88
Text A ………………………………………………………………… 88
Text B ………………………………………………………………… 96
科技文体翻译技巧（五）　数量的译法 …………………………… 101
科技英语摘要写作（五）　摘要的展开句 ………………………… 105
摘要译写示例 ………………………………………………………… 107

Unit 6　Food Engineering ………………………………………… 112
Text A ………………………………………………………………… 112
Text B ………………………………………………………………… 119
科技文体翻译技巧（六）　定义与描述的译法 …………………… 124
科技英语摘要写作（六）　摘要的结束句 ………………………… 128
摘要译写示例 ………………………………………………………… 129

Unit 7　Management ……………………………………………… 133
Text A ………………………………………………………………… 133
Text B ………………………………………………………………… 141
科技文体翻译技巧（七）　实验与比较的译法 …………………… 147
科技英语摘要写作（七）　摘要缩写（Ⅰ） ……………………… 149
摘要译写示例 ………………………………………………………… 150

Unit 8　Agriculture Economy ……………………………………… 154
Text A ………………………………………………………………… 154
Text B ………………………………………………………………… 160
科技文体翻译技巧（八）　标题的译法 …………………………… 164
科技英语摘要写作（八）　摘要缩写（Ⅱ） ……………………… 168
摘要译写示例 ………………………………………………………… 170

主要参考文献 …………………………………………………………… 173

Unit 1　Agronomy

Text A

Transgenic Plant Breeding

　　This area of agronomy involves selective breeding of plants to produce the best crops under various conditions. Plant breeding has increased crop yields and has improved the nutritional value of numerous crops, including corn, soybeans, and wheat. It has also led to the development of new types of plants. For example, a hybrid grain called triticale was produced by crossbreeding rye and wheat. Triticale contains more usable protein than does either rye or wheat. Agronomy has also been instrumental in fruit and vegetable production research. It is understood that the role of agronomist includes seeing whether produce from a field of 'x' meets the following conditions: 1. Land and water access, 2. Commercialization (market), 3. Quality and quantity of inputs, 4. Risk protection (insurance), 5. Agricultural credit.

　　Agricultural production has increased worldwide through the development of varieties with improved yield traits or stress-tolerant characteristics. To this end, crop breeding programs are devised to accumulate alleles from within the gene pool and from cross-compatible wild relatives.

　　Advances in genetic engineering through recombinant DNA technology and spatial and temporal targeted expression of genes facilitate the transfer of precise gene sequences and the transfer of genes across gene pools. Asexual techniques of gene transfer help to engineer new characters that are otherwise very difficult to introduce by conventional breeding. Plants that receive genes using recombinant DNA technology are called genetically modified (GM) crops. A number of GM varieties and hybrids have been developed and cultivated globally.

　　Staple crop systems across broad agroecosystems under a variety of stresses can be addressed by GM crops in conjugation with plant breeding programs. Genetically modified crop varieties can benefit large acreage cropping system in the developed world as well as small farms in the developing world with the support and implementation of complementary national biotechnology policies that safeguard humankind, national trade, and the environment.

The status of transgenic crops has been the subject of focus globally, with several reviews available. Transgenic technology is a supplementary tool to plant breeding and needs to be coupled with classical breeding program for evolving varieties and products. Biotechnology can only add specific and limited traits; therefore, plant breeding must be coupled with transgenic improvement to bring economically viable products to the marketplace.

A number of reports are available on transgenic technology in different crops, but reports on research and development for GM varieties are limited. The authors of this review aim to identify key factors involved in transgenic breeding programs that include event design, event selection, and limitations often encountered in direct adoption of transgenic lines and breeding strategies to overcome the limitations for development of commercial products. They address the dual approach of genetic transformation and conventional plant breeding to develop viable products. Their focus is on the limitations in the introduction of foreign genes to elite genotypes, selection of a candidate transgenic event, and strategies for further breeding program.

Transfer of genes by crossing involves recombination of genetic material, while genetic transformation allows the transfer of only the desired and specific gene sequences. Conventional methods of plant breeding include (Ⅰ) evolutionary methods—germplasm utilization, pedigree methods, and evaluation of genotype × environment interactions and (Ⅱ) revolutionary methods, accelerated introgression of genes and marker-assisted selection.

Development of a new cultivar is a sequential and cyclic process of creating new diversity, recombination, selection for superior recombinants, and testing followed by commercialization. Through transformation, a new genetic variation is created, but it has to be performance-tested like its classically bred counterparts before release and commercialization. In conventional breeding, the breeder selects for desirable recombination among large segregating population, whereas in transgenic breeding, the breeder looks for a defined trait phenotype and then introgresses the transgenic event into a broad range of desirable genetic backgrounds.

Crossing allows natural flow of genes in random combination, and genetic transformation has the advantage of precision transfer of selected DNA sequences. But in the case of transgenics, the site of integration of the transgene is usually random, sometimes resulting in alteration of the DNA sequence to be inserted and sometimes disrupting the recipient genome (although site specific integration is under development).

Genetic engineering is a powerful and useful way to create additional genetic diversity that can be further incorporated into crop breeding programs. In many cases, transgenic events cannot be directly used for cultivation without some breeding because it is important to have the event in the elite backgrounds free of somaclonal variation. Breeding methods like backcrossing can be used to this end.

Evaluation of transgenic expression can be taken up after backcrossing with recurrent nontransgenic parent to realize events with the most potential. Variation among the clones and within the progeny of an event is observed, and hence it is important to screen all the clones rather than dropping a few clones from a single event.

Transformation of parental lines with different genes of interest and pyramiding them in hybrid progeny can be used for commercialization and long-term benefits. With the introduction of more and more transgenic traits into diverse crops, complementary efforts on breeding for biotechnology are essential, with diversion of funds on an equivalent basis, in order not to lose progress. Careful choice of starting material for genetic transformation coupled with precise integration into cultivated genotypes can allow us to reap its full benefits in crop improvement.

(886 words)

New Words

1. transgenic [trænz'dʒenik] *adj.* 转基因的
2. hybrid ['haibrid] *adj.* 杂种（交）的；*n.* 杂交体
3. triticale [ˌtriti'keili] *n.* 黑小麦

4. crossbreed ['krɔs,brid]　*vt.* 使杂交繁育
5. instrumental [,instru'mentəl]　*adj.* 有帮助的
6. stress-tolerant [stres'tɔlərənt]　*adj.* 抗逆性的
7. allele [ə'li:l]　*n.* 等位基因
8. compatible [kəm'pætəbl]　*adj.* 亲和的
9. recombinant [ri'kɔmbinənt]　*adj.* 重组的；*n.* 重组
10. asexual [æ'seksjuəl]　*adj.* 无性生殖的
11. engineer [,endʒi'niə]　*vt.* 转化
12. conjugation [,kɔndʒə'geiʃən]　*n.* 结合
13. complementary [,kɔmpli'mentəri]　*adj.* 补足的
14. supplementary [sʌpli'mentəri]　*adj.* 补充的
15. viable ['vaiəbl]　*adj.* 可行的
16. elite [ei'li:t]　*n.* 精英
17. somaclonal ['səumə'kləunəl]　*adj.* 体细胞克隆的
18. germplasm ['dʒə:mplæzm]　*n.* 种质（生殖细胞传递的遗传物质）
19. pedigree ['pedigri:]　*n.* 家谱
20. cultivar ['kʌltivɑ:]　*n.* 栽培品种
21. segregate ['segrigeit]　*vt.* 使隔离
22. phenotype ['fi:nətaip]　*n.* 表现型，表型
23. introgression [,intrəu'greʃən]　*n.* 基因渗入
24. counterpart ['kauntəpɑ:t]　*n.* 副本，配对物
25. disrupt [dis'rʌpt]　*vt.* 使陷于混乱
26. recipient [ri'sipiənt]　*n.* 接受者
27. genome ['dʒi:nəum]　*n.* 基因（染色体）组
28. backcross ['bæk,krɔs]　*vi.* 回交
29. progeny ['prɔdʒini]　*n.* 后裔
30. equivalent [i'kwivələnt]　*adj.* 等价的；*n.* 等价物

Useful Expressions

1. selective breeding　选育
2. gene pool　基因库
3. wild relative　野生亲缘
4. recombinant DNA technology　DNA 重组技术

5. expression of gene   基因表达
6. gene sequence   基因序列
7. transfer of gene   基因转换（移）
8. asexual technique   无性繁殖技术
9. genetically modified crop   转基因作物
10. transgenic technology   转基因技术
11. transgenic line   转基因株系
12. germplasm utilization   种质资源利用
13. pedigree method   系谱法
14. sequential and cyclic process   顺序和循环过程
15. segregating population   分离群体
16. transgenic event   转基因片段
17. somaclonal variation   体细胞克隆（无性系）变异；体细胞突变
18. recurrent nontransgenic parent   非转基因轮回亲本
19. parental line   亲代系
20. cultivated genotype   改良的基因型

Notes

1. 本文选自美国农学会（American Society of Agronomy，ASA）官网（https：//www.agronomy.org）和美国作物科学协会（Crop Science Society of America，CSSA）官网（https://www.crops.org/about‐crops）。文章在线出版于 2009 年 8 月 7 日（Crop Sci，2009，49：1555-1563，Copyright 2009 Crop Science Society of America，677 S. Segoe Rd.，Madison，WI 53711 USA）。

2. 转基因作物（genetically modified crops，GMC），是利用基因工程将原有作物的基因加入其他生物的遗传物质，并将不良基因移除，从而造成品质更好的作物。通常，转基因作物具有可增加作物的产量，改善品质，提高抗旱、抗寒和其他特性。但反对者表示，对转基因食物进行的安全性研究都是短期的，无法有效评估人类进食转基因食物几十年后或者更久以后的风险；转基因生物不是自然界原有的品种，对于地球生态系统来说是"外来生物"，可能导致传统生物的基因污染。

3. 转基因技术（genetically modified technology，GMT）源于进化论衍生的分子生物学，基因片段来源可以是提取特定生物体基因组中所需要的目的基因，也可是人工合成指定序列的 DNA 片段。DNA 片段被转入特定生物中，与其本身基因组进行重组，再从重组体中进行数代人工选育，从而获得具有稳定表现特定的遗传性状的个体。该技术可以使重组生物增加人们所期望的新性状，培育出新品种。常说的"遗传工程、基因工程、遗传转化"均为转基因同义词。

Exercises

Part Ⅰ Vocabulary and Structure

Section A Match each word with its Chinese equivalent.

1. nitrogen   A. 酶
2. cultivation   B. 除草剂
3. phytochemical   C. 淀粉
4. fermentation   D. 豆类，豆荚
5. anaerobic   E. 氮
6. forage   F. 厌氧的
7. starch   G. 发酵
8. herbicide   H. 植物营养素
9. legume   I. 栽培
10. enzyme   J. 饲料

Section B Fill in the blanks with the words or expressions given below. Change the form where necessary.

| recurrent | backcross | evolution | couple | counterpart |
| instrumental | stress-tolerant | recombinant | hybrid | conjugate |

1. In the production of grains, fruits and vegetables, the agronomy research plays a (an) _____ role.
2. Pest-resistant and _____ crops can be developed to reduce the risk of crop failure due to drought and disease.
3. _____ gene technology is widely employed in research and development for strain improvement.

4. All these brightly colored _____ are so lovely in the garden.
5. For bacteria, they are binary (二元的) fission (分裂生殖法) and _____.
6. Each of them has a _____ column in the database tables.
7. Plant breeding must be _____ with transgenic improvement to bring economically viable (可行的) products to the marketplace.
8. Crops are derived from the _____ process as all other living creatures, from the wild to the domestic.
9. _____ breeding is a useful method to transfer favorable alleles from a donor parent into a recipient parent.
10. At the conclusion of the _____ run, you should gather and evaluate the utilization levels for all four resources of crops.

Part II Translation

Section A Translate the following sentences into Chinese.

1. This area of agronomy involves selective breeding of plants to produce the best crops under various conditions.
2. The large acreage cropping of genetically modified crop varieties can benefit the developed world as well as small farms in the developing world.
3. In conventional breeding, the breeder selects for desirable recombination among large segregating population.
4. In transgenic breeding, the breeder looks for a defined trait phenotype and then introgresses the transgenic event into a broad range of desirable genetic backgrounds.
5. Evaluation of transgenic expression can be taken up after backcrossing with recurrent nontransgenic parent to realize events with the most potential.

Section B Translate the following sentences into English.

1. 减产可能是由于化肥缺乏、耕作技术落后和在关键时间水分过多或过少。
2. 如果低地必须用于种植豆类作物，就应翻地做垄（ridge）。
3. 饲料作物是专为满足牲畜营养需要而种植的作物，可以包括谷类作物、油料作物和牧草作物。
4. 目前大多数转基因工程中，人们使用组成型启动子（constitutive promoter）

驱动外源基因表达（alien gene expression）。
5. 在美国，主要的农作物包括玉米、大豆、小麦、干草、水稻、花生和棉花。

Text B

Crop Science

The Future of Agriculture

The evolution of agriculture within the last 11,000 years marked the first major *inflection point*（转折点）in food yield and changed forever the character of the human condition. The application of technology to agriculture early in the 20th century induced the next major crop yield inflection point. Identifying the technological wellspring from which increased rates of productivity will be obtained in the decades ahead is far less obvious than during the last century.

The agronomic challenge for the decades to come is to increase productivity per unit of land enough to *preclude*（排除）*appropriation*（挪用）of other ecosystems for cropland expansion while simultaneously increasing the efficiency of production inputs, reducing their *leakage*（泄露）to the environment, and sustaining the integrity of those ecological processes that *undergird*（加强）these intense biological production systems. This excerpt from the abstract of an article by Fred P. Miller, retired Professor of Soil Science at the School of Environmental and Natural Resources at The Ohio State University was written in celebration of 100 years of The American Society of Agronomy.

Every Day, Crop Science

Every day, everyone is impacted by crop science. From the endless green fields of corn and soybeans which cover the Midwest, the vibrant yellows of sunflowers in Canada, the expansive rice paddies of Asia, the vast acres of cotton drying under the hot Southwestern sun, to the lush green mountains of coffee growing in Central America, these crops do not just happen. Hard work on the part of the grower, aided by the crop sciences makes these crops possible.

Crop scientists are at the *intersection*（交叉点）of plant and soil sciences and work to improve crops and agricultural productivity while effectively managing pests and weeds. They make this possible through the application of soil and plant sciences to crop production that incorporates the wise use of natural resources and conservation practices to produce food, feed, fuel, fiber, and pharmaceutical crops while maintaining and improving the environment.

A Day in the Life of a Crop Scientist

A career in crop science keeps you in the center of efforts to increase the production of food, feed, fuels and fiber, for a growing world *citizenry*（公民）. The crop scientist has many career paths. You'll find agronomists working in research, teaching and extension at colleges and universities, for the USDA at their Agricultural Research Stations, in extension offices, for companies, and as consultants in agribusiness. Interested in a career in crop science? Discover more with our career brochures, and view the list of colleges and universities with courses and programs in agronomy, crop science, soil sciences, and related *disciplines*（学科）.

The Science of Crops

The evolution and ongoing development of agriculture, enabled by science, is the focus of agronomists and crop scientists. Scientific research to enhance productivity while sustaining the integrity of ecological processes encompasses crop science, soil sciences, and environmental science. The research is communicated and transferred among agronomists and those in related fields on topics of local, regional, national, and international significance. This research may then be used for practical applications. Scientific articles on specific research are available 18 months after publication and presentations from Annual Meetings are available one year after presentation.

Agronomic Crops

Agronomic crops are grown in all fifty states and many countries around the globe. What is an agronomic crop? Agronomic crops typically involve a crop that is grown for grain, feed, or for processing into oil, starch, protein and flour. Major agronomic crops in the US include corn (grown for feed, *ethanol*（酒精）or processing), soybeans, wheat, hay (*alfalfa*（紫花苜蓿）and *legume*（豆科植物）and grass mixtures), rice, peanuts and cotton. Hay is also considered a forage. Growing agronomic crops is an integrated system. It is important to understand how the soil works and interacts with the growing crop, what nutrients the growing crop needs and when and how these nutrients can be applied, how a crop grows and how the environment interacts with the crop at all growth stages. In addition, it is important to know how pests (weeds, insects and diseases) affect crops at various growths and how to control crop pests. Growing crops involves soil, plant, crop and weed sciences, plant genetics, *entomology*（昆虫学）and plant *pathology*（病理学）.

Crop production, as a result of scientific and applied research, continually changes. Just in the last decade many producers and crops retailers have started using *global positioning*（全球定位）to gain more understanding of how soil fertility varies throughout a field as well as to apply fertilizers based on how this fertility varies. Global positioning is also used to measure the size of a field, measure yield at a given place in the field, and guide implements across the field to prevent overlap and improve land use efficiency. Crop genetics has also changed tremendously. Producers are able to grow crops that are resistant to certain environmentally safe herbicides. They are also able to grow crops that are resistant to injurious insects allowing producers to eliminate or reduce overall insecticide use. Examples of future crop genetics improvements include growing crops for the healthier oil that can be extracted, for certain starch characteristics, and growing grass crops that can yield more with less nitrogen and that will produce well even during seasons when rainfall is limited. Scientific research provides us with valuable information on how to efficiently and effectively grow agronomic crops.

Feed Crops

Feed crops are crops grown specifically to meet livestock nutritional needs. They may include grain, oilseed, and forage crops. Grain crops are grasses that are grown for their dry, edible seeds. These include small grains such as wheat, oats, barley, and rice and larger, taller crops like corn and sorghum. Oilseed crops are those with seeds high in oil and protein. A commonly grown oilseed crop for animal feed is soybean.

Forage crops are livestock feeds grown for their edible plant parts other than the separated grain. These parts typically are the stems and leaves of green, actively growing grasses and legumes. Livestock, particularly *ruminants*（反刍动物）（cattle, sheep）, may consume forages within the field as grazed pasture or they may be fed them as stored forages. Stored forages include hay, *silage*（青贮饲料）, and *green chop*（青刈饲料）.

Hay crops are forages that are cut while still green, allowed to dry in the field, processed and then stored before being fed to livestock. Silage crops are forages that are harvested in a green, *succulent*（多汁的）condition and stored under *anaerobic*（厌氧的）conditions where controlled fermentation breaks down plant sugars to organic acids, especially *lactic acid*（乳酸）. Green chop refers to forages that are cut, harvested, and fed while still in a green and wet condition.

(1,095 words)

Comprehension of the Text

Choose the best answer to each of the following questions.

1. Within the last 11,000 years, what did the evolution of agriculture influence?
   A. The process of human evolution.
   B. The civilization of mankind society.
   C. The application of technology.
   D. The character of the human condition.
2. What is considered as the agronomic challenge in the future decades?
   A. To preclude appropriation of other ecosystems for cropland expansion.
   B. To improve the overall productivity of grains.
   C. To increase enough productivity per unit of land.
   D. To sustain the integrity of those ecological processes.
3. Which of the following is the main task of crop scientists?
   A. To manage plant and soil in field.
   B. To improve crops and agricultural productivity.
   C. To deal with pests and weeds with farmers.
   D. To train peasants in productivity.
4. For growing population, which of the following is the center of efforts in crop science career?
   A. To increase the production of food, feed, fuels and fiber.
   B. To improve the productivity of grains, vegetables in general.

C. To promote the production of agriculture products.

D. To proceed the productivity of agronomic products.

5. To enhance productivity while sustaining the integrity of ecological processes, what disciplines do scientific research include?

   A. Crop science, soil science and environmental science.

   B. Crop science, environmental science and cultivation science.

   C. Crop science, climate science and environmental science.

   D. Crop science, environmental science and agronomic science.

6. What are the major agronomic crops in the US?

   A. Maize, rice, soybeans, wheat, peanuts, cotton and hay.

   B. Corn, wheat, soybeans, peanuts, alfalfa and legume grass mixtures.

   C. Corn, rice, beans, wheat, peanuts, cotton and hay.

   D. Maize, soybeans, wheat, hay, rice and cotton.

7. Which of the following practices involves soil, plant, crop and weed sciences, plant genetics, entomology and plant pathology?

   A. Planting seedlings.   B. Breeding sprouts.

   C. Cultivating grains.   D. Growing crops.

8. When it comes to crop genetics, what can be regarded as one of the good traits of crops for producers?

   A. Resistance to injurious disease.

   B. Resistance to overall disease.

   C. Resistance to injurious insects.

   D. Resistance to injurious herbs.

9. What are feed crops grown specifically to meet?

   A. Husbandry living conditions.

   B. Livestock nutritional needs.

   C. Poultry growth demands.

   D. Animal propagation requirements.

10. What kind of forage crops are harvested in a green, succulent condition and stored under anaerobic conditions ?

    A. Silage forages.

    B. Fermentative crops.

    C. Anaerobic forages.

    D. Green chops.

科技文体翻译技巧（一）

翻译的增减

在英汉互译时，有时采取增加或减少词语、结构的翻译方法，即增译法和减译法。该翻译方法的作用和意义在于在忠实原文内容前提下，使译文语法结构、语言意义或修辞效果更加完整、通顺和明确。需要指出的是，增译法和减译法二者是相互性的关系，即汉译英时使用增译法，英译汉时则须为减译法，反之亦然。

一、逻辑关系（结构）中关联词的增译和减译

一般文体中，汉语逻辑关系通常以上下文语境表达，而英语逻辑关系主要以关联词表达。汉译英时多需增补出关联词，英译汉时可少用甚至不用关联词，因此在译文中要格外注意关联词的增补和减少使用。在科技文体中，逻辑关系的表达除注意使用关联词外还需格外注意其结构顺序的调整变化，此时增译和减译就要格外慎重。

【例】1. 不同时期栽培的水果，由于花期所经历的气候条件不同，其坐果率也各有不同。

The percentage of fruit setting is different on different planting dates owing to the fact that climatic conditions are different during flowering season.

2. It is possible for cow, deer, goats, pigs and sheep to get the disease of foot and mouth. Though many of them can recover, the disease will weaken these affected animals which will have cuts in their mouths and on their feet.

牛、鹿、山羊、猪和绵羊都会感染口蹄疫。受感染的牲畜会在嘴和蹄上出现伤口，很多受到感染的牲畜可以康复，但疾病使它们变得瘦弱。

二、逻辑关系中句子成分（主语、谓语、宾语、定语、状语和补语）的增译和减译

（一）句子成分的增减翻译

文中句子成分搭配的合理性并非在译文中一定显得合理，因此对于译文而言，在忠实原文的同时一定要关注句子成分的增加或减少，保持译文语言成分的完整性。在科技文体中，各种句子成分决定着科技信息的完整性、明确性、

准确性、结构的严谨性和数据的可靠性等。

【例】 1. Air, food, water and heat are four requirements of all living things.
空气、食物、水和热量是一切生物赖以生存的四个条件。

2. The earthworms use as food some of the plant material in the soil they swallow.
蚯蚓以土壤中的某些植物为食。

3. This is a stable variety of common sorghum, and another one is an addition line developed by researchers from a cross between common sorghum and some weed.
这是一个稳定的普通高粱品种，而另一个则是高粱与某种草的杂交后代。

4. The solution shall be standardized on the day it is used.
溶液应该在使用当日进行标定。

（二）主语成分的特殊地位及其增译和减译

需特别指出，汉语无主句较多，在汉语表达中可以省略主语变成无主句，甚至可以将谓语动词也省略，所以在汉译英时必须格外关注汉语中隐含的逻辑主语，多需在英文中增补翻译。反之，英译汉时汉语中多数主语不必表达出来，所以英译汉时，需特别注意英文中主语可能在汉语中表现得多余，多需在汉语中减少翻译。应该指出，科技文体行文时各种逻辑成分必须谨慎、明确地增减译。

【例】 1. 要是能够看到转基因试验成功，那该有多好啊！
If only we could see the success of the transgenic experiment!

2. 针对我国丰富的畜种资源，研究人员应迅速进行研究并加以利用。
The rich resources of animal breed of China should be promptly studied and utilized.

3. Indeed, the reverse is true.
事实情况则恰恰相反。

4. What about fertilizing the crops right away?
立刻给作物施肥，你觉得如何？

三、汉语短语结构的增译和减译

（一）汉语"抽象名词＋范畴词"的增减译

科技文体中的汉译英，通常英文只保留汉语"抽象名词"而省略"范畴词"（例如现象、状态、工作、情况、方法、问题、方面、关系、过程、作用、程度、因素、值、量、率等）。

【例】1. 植物生长的因素包括内因和外因等诸多方面的要素。
　　　There are internal and external factors which affect plant growth.
　　2. 化肥与农药的使用必须限制在一定的范围之内。
　　　The amount of fertilizer and pesticide must be limited.

（二）汉语中"结构动词＋动作名词"的增减译

科技文体中的汉译英，通常英文只保留汉语"动作名词"而省略"结构动词"（例如加以、进行、给予、予以、做了、发挥、体现、表达、有所等）。

【例】1. 在组织和细胞培养过程中容易发生染色体数目和结构的变异。
　　　The number and structure of chromosome were easily varied during tissue and cell culture.
　　2. 对腐殖质酸钠对植物吸收磷的影响加以分析
　　　Analysis of effects of sodium humate on the uptake of phosphorus by plants

四、虚词中具有特殊地位和功能的词的增译和减译

（一）介词和冠词的增译和减译

这两种虚词在科技文体翻译和写作中具有十分特殊的功能：第一、汉语没有冠词，英语通常离不开冠词，汉译英时大多数地方需要增译冠词，反之则需要减译冠词；第二、汉语使用介词的频率没有英语高，尤其是在科技文体的标题中英语的介词使用几乎可以完全代替谓语动词的使用。

【例】1. 这篇论文的题目是《寒冷药与温热药配伍初探》。
　　　The title of the paper is *Preliminary Study on the Combined Use of Drugs of Hot and Cold Nature*.
　　2. 在1千克/组下使用秋水仙碱20毫克，则生长极好。
　　　In the treatment of 20 mg colchicines in 1 kg of one group, the plants showed excellent vigor in growth.
　　3. Science and Technology for the Development of Agriculture Report of UN Conference on the Application of Science and Technology for the Benefit of the Less Developed Areas
　　　在为欠发达地区利益召开的联合国科技应用会议上所做的农业发展科学技术报告

（二）虚词中数量词的增减译

科技文体中数量的表达频率很高，汉语中许多数量词（例如许多、很多、大量、相当等），英文中可用复数表达，即减译；反之在英译汉中需增译。

【例】1. 过去的一些研究工作曾证明，信使核糖核酸（mRNA）可以调控遗传性状的转化。

Previous studies have shown that mRNA mediated the transformation of genetic characters.

2. 研究人员在含低浓度 GA（1~4 毫克/千克）的 MT 培养基上只能得到球形或心形胚状体。

Only globular or heart-shaped embryoids were obtained on MT medium containing GA at low concentration (1-4 mg/kg).

五、语境中词句特殊含义的增译和减译

在科技文体中，特定语境（例如概括性、注释性和暗示性等情况）和特定专业下，许多实词、虚词、短语、句子和逻辑结构（关系）等意义需要在中英文互译时做增译或减译。

【例】1. 全世界在科学工作中都使用十进位制，甚至那些本国计量制基于其他标准的国家也如此。

The decimal system is used for scientific purpose throughout the world, even in countries whose national system of weights and measurements are based on other scales.

2. 调查发现酸菜为东北人喜食的副食品之一，而酸菜普遍为一种真菌——白地霉所污染。

The investigation finds that Northeast people are fond of eating a kind of pickled cabbage, which is often contaminated with the fungus *Geotrichum candidum* Link.

科技英语摘要写作（一）

信息型摘要

一、摘要概述

摘要是学术论文的重要组成部分。摘要不仅能概括文章主要内容，而且有助于审稿人审稿，还利于读者迅速判断文献的主要内容，更能为数据库进行文

献检索提供便利。摘要内容，一般包括3大基本部分：①提出研究的主要目的和范围，说明本文是研究什么的或本文要解决什么问题，有两个要求，一是不要提及背景知识或尽可能少提，二是第一个句子尽量避免重复标题或部分标题；②描述研究的内容、过程、方法、条件和主要的仪器设备，所有这些都需跟公式、实验、图表等一起详细描述；③总结实验结果和主要结论，旨在使实验过程中得出的结果或图表、曲线等使得观点更清晰、更有说服力。除以上3个基本部分外，摘要最后部分通常将结果与其他最新研究结果相比较，以强调这些结果的贡献性及创造性。摘要一般分为3种：信息型摘要、指示型摘要和指示-信息型摘要。

本书所涉及的论文摘要涵盖了农学、中药、生命、资源、动物科学、食品工程、信息工程和农业经济管理等方向，学科间摘要的语言存在差异性，其表达也各具特点。

二、信息型摘要

一般说来，信息型摘要在所有科技论文摘要中占有绝大部分的比例，是最常见的。它揭示出论文传递的主要信息甚至详细新信息，叙述问题、领域、方法及主要的结论、建议。信息型摘要一般包括文章某些重要内容的梗概，主要说明作者写此文章的目的，或者文章主要想解决的问题；作者主要的工作过程、所用的方法和主要设备和仪器；作者通过此工作过程最后得到的结果和结论等，有时也顺便提及所得结果、结论的应用范围和应用情况。信息型摘要多用于科技杂志或科技期刊的文章，也用于会议记录中的会议论文，以及各种专题技术报告等。信息型摘要的主要部分包括：①问题陈述，引出研究课题，或提出问题，经常用一至两句话。陈述部分不得忽略，这是因为问题所在实际上只有作者本人知道。②重大发现或新方法，一般表述论文中至关重要的信息，给出研究结果、结论、建议，以及对深入研究的暗示，但有时摘要也提供其他信息，例如许多技术性课题研究获取某些结果的新的或不寻常的方法，而这些结果之前已经由其他方法取得过，此时摘要就主要介绍研究方法而不是结果。

【例】 Abstract: Chinese advantages in soybean production and trade have disappeared. China is the biggest country of soybean import now. The reasons which caused this phenomenon are as follows: domestic demand of soybean increased rapidly; the price was lower and the quality was better in the international market; we have made the protective policies excessively. The countermeasures are as follows: restructure agriculture industry quickly; producing soybean with advanced technology; develop soybean processing industry, and making full use of the advantage that we produce non transgenic soybean.

三、摘要翻译

一般而言，英文摘要是中文摘要的转译，所以应按照英文的语言习惯准确地译出中文摘要的内容。①就篇幅而言，字数没有硬性规定，但普遍认为以150～180个单词为宜。②就时态而言，英文摘要的时态运用也以简练为佳，常用一般现在时、一般过去时、现在完成时和过去完成时，进行时态和其他复合时态基本不用。一般现在时主要用于陈述性、资料性摘要中。一般过去时主要用于说明某个具体项目的发展情况，介绍技术研究项目的具体资料。现在完成时是把过去发生的或过去已完成的事情与现在联系起来，说明研究的发展背景，介绍已结束的研究。③就语态而言，科技论文主要是讲述客观现象和真理，摘要更是对论文内容的高度概括，因而普遍采用的是被动语态。采用被动语态可以避免提及动作的执行者，使行文显得客观。同时，被动句在结构上有较大的调节余地，有利于扩展名词短语，扩大句子的信息量，有利于突出有关概念、问题、事实、结论等内容。④就句式而言，多用长句、复合句。汉语的句法特征是意合，句子之间意连形不连，句子之间的意义及关系隐含其中。而英语的句法特征是形合，句子以形连表示意连，表示关系的关联词语（如关系代词、关系副词）都起重要的纽带作用。

摘要译写示例

示例一

【摘要中文原文】

摘要：本研究是利用 hrpZPsg12 转化玉米，以期得到对玉米病害具有广谱抗性的新种质材料，为抗病育种提供新的种质资源。以植物表达载体 pCAMBIA3301 为基础载体，采用 PCR 法从克隆载体 pGM-hrpZPsg12 克隆来自丁香假单胞菌大豆致病变种（P. syringae pv. glycinea）的 hrpZPsg12 基因，两端分别引入 Xho I 酶切位点，构建了 1 个具有卡那霉素和除草剂草丁膦抗性标记的植物表达载体 pCAMBIA3301-hrpZPsg12，并通过热激法转化入大肠杆菌 DH5α 中。采用花粉管通道法将构建好的植物表达载体的重组质粒导入玉米优良自交系综 31 中。结果表明，对得到的 575 株转化植株经 PCR 检测有 45 株呈阳性，阳性转化率为 7.82％。

【摘要原版译文】

Abstract：The objective of this study is to obtain new maize germplasm materials being of broad-spectrum resistance to maize diseases and provide a new resources for maize resistance breeding by transferring hrpZPsg12 gene into good inhybrid line Zong 31. HrpZPsg12 gene, which from *Pseudomonas syringae* pv. *glycinea*, was cloned from cloning vector pGM-hrpZPsg12 by PCR method and was transferred into the based plant expression vector pCAMBIA3301…. The recombinant plasmids of pCAMBIA3301-hrpZPsg12 were extracted from constructed DH5α and were imported into maize by pollen tube pathway method.

【摘要修改译文】

Abstract：The objective of this study is to obtain new maize germplasm materials of broad-spectrum resistance to maize diseases and provide new germplasm resources for maize disease resistance breeding by transferring hrpZPsg12 gene into good maize inbred line Zong 31. HrpZPsg12 gene, from *Pseudomonas syringae* pv. *glycinea*, was cloned from cloning vector pGM-hrpZPsg12 by PCR method and transferred into the based plant expression vector pCAMBIA3301…. The recombinant plasmids of pCAMBIA3301-hrpZPsg12 were extracted from constructed DH5α and imported into maize Zong 31 by pollen tube pathway method. The results showed that 45 positive plants were detected from the obtained 575 transgenic plants by PCR, and the positive conversion rate was 7.82％.

【主要修改意见】

1. 正确判断逻辑关系。将原版译文中的"being of"修改为"of"，表示从属关系，说明事物的特性。
2. 正确表达单复数概念。如"a new resources"，应去掉"a"。
3. 避免遗漏信息。第一句漏译信息"disease"。
4. "which"引导的非限定性定语从句表述有缺失，缺少谓语动词。建议将"which from"修改为"from"，直接使用介词短语，使行文更加简洁。
5. 避免相同、相似句法结构的重复翻译。如"was cloned from … and was transferred into"的后半部分翻译可以修改为"and transferred into"；另如

"were extracted from ... and were imported into"的后半部分翻译可以修改为"and imported into"。

示例二

【摘要中文原文】

摘要：本研究旨在建立一套适合山葡萄资源的果实品质评价方法，并为优质山葡萄资源快速筛选及品种选育提供理论参考。根据果实主要内在品质的差异，应用主成分分析法和聚类分析法对对果汁颜色不同的14份山葡萄资源进行综合评价和分类。结果表明，前3个主成分的累积方差贡献率为89.09%；第一主成分主要由单宁、可溶性糖、糖酸比和白藜芦醇决定；第二主成分主要由原花青素和总黄酮决定；第三主成分主要由可滴定酸决定。基于果实品质综合分值，14份山葡萄资源聚成3类，主成分分析法构建的综合评价体系揭示了果汁颜色不同的山葡萄资源果实品质的差异。

【摘要原版译文】

Abstract: In order to establish a suitable fruit internal quality comprehensive evaluation system and provide a theoretical basis for high quality amur grape resources of quick selecting and breeding, fourteen resources were measured. Advanced selection and classification of fourteen resources were carried out by principal component analysis and cluster analysis according to the differences in main quality characteristics. The results showed that the cumulative proportion of the former three principal components was 89.09%, the first principal component was highly connected with tannins, soluble sugar, sugar-acid rate and resveratrol; determining the size of the second principal component was mainly procyanidins and total flavonoids; titratable acid was an important contributor to the variance for the third principal component. Therefore, the fourteen resources were divided into 3 groups based on their comprehensive scores. From the above results, the comprehensive evaluation system built in this study could be used to reveal the quality diversity of *Vitis amurensis* resources from different color units.

【摘要修改译文】

Abstract: In order to establish a suitable fruit internal quality comprehensive evaluation system and provide a theoretical basis for high quality amur grape resources of quick selecting and breeding, fourteen resources were measured. Advanced selection and classification of fourteen resources were carried out by principal component analysis and cluster analysis according to the differences in main quality characteristics. The results showed that the cumulative proportion of the former three principal components was 89.09%; the first principal component was mainly determined by tannins, soluble sugar, sugar-acid rate and resveratrol; the second principal component was mainly determined by procyanidins and total flavonoids; titratable acid was an important contributor to the variance for the third principal component. Therefore, the fourteen resources were divided into 3 groups based on their comprehensive scores. From the above results, the comprehensive evaluation system built in this study could be used to reveal the quality diversity of *Vitis amurensis* resources from different color units.

【主要修改意见】

1. 正确使用标点断句。原版译文"The results showed that ... 89.09%, the first principal component was..."中的逗号不能连接前后两个句子，应使用分号隔离。
2. 科技文体用词准确化。原版译文"highly connected with"不能准确表达"主要由……决定"，应译为"mainly determined by"。
3. 通顺表达语序。原版译文"determining the size of the second principal component was mainly..."的语序需要调整，以正确表达逻辑关系。应译为"the second principal component was mainly determined by..."。

Unit 2   Traditional Chinese Medicine

Text A

Harmonizing Traditional Chinese and Modern Western Medicine: A Perspective from the US

The current interest in traditional and complementary medicine in the United States is attracting attention in many parts of the community—the health care industry, governmental agencies, media and the public. An increasing number of insurers and managed care organizations are providing benefits for traditional medicine, a majority of U. S. medical schools now offer courses covering traditional medicine, and, as Eisenberg's national studies have revealed, more people are using complementary therapies. To facilitate research on the effectiveness of alternative therapies, the National Center for Complementary and Alternative Medicine (NCCAM) received a budget of $50 million in 1999. Recognizing the need to encourage quality and quantity of scientific information on botanicals, as well as develop a systematic evaluation of safety and efficacy of dietary supplements, two research centers were also established this year to investigate the biological effects of botanicals.

Many patients are using traditional and modern medical paradigms concurrently, creating a need for the appropriate and smooth merger of the two medicines. The theories and techniques of traditional Chinese medicine (TCM) encompass most practices classified as complementary medicine in the United States, and have become increasingly important in the health care system. Traditional Chinese medicine is affordable, low-tech, safe and effective when used appropriately. Ongoing research around the world on acupuncture, herbs, massage and Tai-Chi has shed light on some of the theories and practices of TCM. Evidence derived from vigorous research design as well as patient demand are fueling the merger of TCM with modern medicine at the clinical level, while more academic researchers and institutions are becoming more interested in the potential of integrating these two healing traditions.

Acupuncture

Based on evidence reviewed during the 1997 NIH Consensus Conference, the NIH Consensus Development Panel conservatively recommended that acupuncture may be used as an adjunct treatment, an alternative, or part of a comprehensive management program for a number of conditions. The panel ascertained that acupuncture can be used to treat post-operative and chemotherapy induced nausea and vomiting, as well as post-operative dental pain. It was also recommended as an adjunct treatment or an acceptable alternative for addiction, stroke rehabilitation, headache, menstrual cramps, tennis elbow, fibromyalgia, myofacial pain, osteoarthritis, low back pain, carpal tunnel syndrome, and asthma.

Future clinical trials that test acupuncture within the framework of traditional Chinese medicine are likely to provide a more appropriate and clinically meaningful assessment of acupuncture efficacy than the current generation of clinical trials which use a diagnosis framed primarily in biomedical terms. The scientific rigor of current research must continue; however, the NIH approach towards data analysis is too strict and limits potentially useful indications. Unlike drugs, acupuncture is more akin to surgery and physical therapy in terms of therapeutic modalities. Hence, the evaluation of evidence for efficacy in acupuncture ought to be similar to these therapeutic interventions. For the time being, evidence based on large case series should be considered in determining recommendations for clinical practice while evidence derived from more vigorous research designs are being carried out.

In elucidating the mechanisms of acupuncture and exploring its role in a variety of situations, innovative techniques are beginning to be utilized. Studies on acupuncture in terms of its neuroanatomic and neurophysiological bases, bioelectrical properties, analgesia effects, and its role in regulation in areas such as gastrointestinal, immunological and cardiovascular functions are being carried out. More intense research with increased funding and scientific vigor, in and out of the US, will likely uncover additional areas where acupuncture may prove useful. This will further drive the adoption of acupuncture as a common therapeutic modality, not only in treatment, but also in prevention of disease and promotion of wellness. With technological advancement, innovative methods of acupuncture point stimulation will continue to be explored and perfected. Basic research on acupuncture will also help facilitate improved understanding of neuroscience and other aspects of human physiology and function.

Because of heightened patient demand and better understanding of the role of acupuncture in health care through research and clinical experience, the biomedical establishment, health insurance industries, physicians and other health care providers are beginning to take an interest in acupuncture. In time, those who do not embrace acupuncture will be at a disadvantage. As the efficacy and cost saving potential of acupuncture is more widely recognized, there will be an even stronger push for more insurance companies, medical groups, and even Medicare to provide coverage of acupuncture treatment. We will witness acupuncture being utilized increasingly in outpatient settings, hospitals, rehabilitation units and hospices. An increasing number of physician acupuncturists as well as non-physician acupuncturists are working in different clinical settings. Some licensed acupuncturist specialists will work in specialized areas.

In the new millennium, the practice of acupuncture will be guided not only by traditional Chinese medicine concepts, but also by data generated through research advances in diverse fields such as neuroscience, molecular biology, chronobiology, computer and information science, energetics, integrative physiology and innovative clinical trial methodology.

Herbs

Humans and animals have tested and used botanicals to relieve their suffering since ancient times. The appropriate use of Chinese herbs requires proper TCM diagnosis of the zheng(pathophysiological pattern) of the patient, correct selection of the corresponding therapeutic strategies and principles that guide the choice of herbs and herbal formulas. When appropriately prepared and used, herbs can be safe and effective. However, when used without proper guidance, a wide array of complications may result.

Modern scientific investigations on plant-based medicine have been carried out in many parts of the world, including clinical trials of botanical combination products. Clinical research methodologists should take the theoretical construct and clinical approach of TCM into consideration when designing trials. Research designs such as randomized controlled trials (RCTs) have advantages and disadvantages in determining the efficacy of any therapeutic intervention, and can be carried out for botanicals, as seen by a study on herbal formulas for irritable bowel syndrome. Yet, we should seek approaches other than conducting a clinical trial for each product to evaluate safety and efficacy. Alternatives to RCTs include quasi-experiments, cohort studies, case-control studies, and "N = 1" trials. These methods have their advantages and limitations but may be more suited to the evaluation of herbal efficacy. The accurate measures of patient-centered outcomes both generic and disease-specific are important regardless of the design of the study. Above all, the appropriate study design depends on the research question and hypothesis being tested.

Evaluating evidence is both difficult and subjective. The synthesis of evidence is completely dependent on the completeness of the literature search, which is often not available for foreign studies, as well as the accuracy of evaluation. Also, there are situations when neither RCTs nor database analyses separately can answer the question of interest due to different populations being used in the various kinds of studies. Consensus in the real world of health care often requires using information that is less stringent than so-called hard data. Realizing this, we should recognize the research and practice of herbal therapies in China, Korea and Japan when making recommendations for clinical practice. The pharmacological basis for many herbs has been determined in these studies and, as long as safety is assured, their findings should be considered when making recommendations. It is essential that researchers and practitioners be educated in both traditional and western medicines in order to perform research appropriately and treat patients effectively.

Integrative East-west Medicine

Harmonizing traditional medicine and modern medicine is more than utilizing modern research design or scientific technology to assess traditional medicine; it should include assessment of the intrinsic value of traditional medicine in society. Political, economic and social factors play as equally an important role as research and education in the eventual blending of the two healing traditions.

On the clinical level, blending involves the integration of the concepts and techniques of the two systems—modern medicine's analytical, quantitative, mechanistic approach with the systemic, holistic, individualistic approach of TCM. This framework is applied through the process of diagnosis, prevention, treatment, and rehabilitation and guides the use of the appropriate techniques, allowing the strengths of TCM to compensate for the weaknesses of modern western medicine. As our graying society falls victim to an increasing number of chronic illnesses, we need a health paradigm that solves problems and provides affordable, effective health care for all. We believe that integrative east-west medicine is a candidate for such a model of medicine.

(1,403 words)

New Words

1. harmonize ['hɑːmənaiz]　*vt.* 使和谐
2. therapy ['θerəpi]　*n.* 疗法
3. botanical [bə'tænikəl]　*n.* 植物性药材
4. efficacy ['efikəsi]　*n.* 功效
5. dietary ['daiətəri]　*adj.* 规定饮食的
6. paradigm ['pærədaim]　*n.* 范例
7. concurrently [kən'kʌrəntli]　*adv.* 同时发生地
8. merger ['məːdʒə]　*n.* 融合
9. encompass [in'kʌmpəs]　*vt.* 包含；包围
10. acupuncture ['ækjupʌŋktʃə]　*n.* 针刺疗法
11. massage ['mæsɑːʒ]　*n.* 按摩
12. consensus [kən'sensəs]　*n.* 一致
13. adjunct ['ædʒʌŋkt]　*adj.* 附属的
14. panel ['pænl]　*n.* 专家组
15. ascertain [ˌæsə'tein]　*vt.* 确定
16. chemotherapy [ˌkeməu'θerəpi]　*n.* 化学疗法
17. nausea ['nɔːzjə]　*n.* 恶心
18. vomit ['vɔmit]　*vi.* 呕吐

Unit 2　Traditional Chinese Medicine

19. rehabilitation ['ri:hə,bili'teiʃən]　*n.* 康复
20. fibromyalgia [,faibrəumai'ældʒi:ə]　*n.* 纤维性肌痛
21. myofacial [miəu'feiʃiəl]　*adj.* 面部肌肉的
22. osteoarthritis [,ɔstiəuɑ:'θraitis]　*n.* 骨关节炎
23. asthma ['æsmə]　*n.* 哮喘，气喘
24. rigor ['rigə]　*n.* 严密
25. akin [ə'kin]　*adj.* 类似的
26. modality [məu'dæliti]　*n.* 形式
27. elucidate [i'lusideit]　*vt.* 阐明
28. neuroanatomic [,nju:rəuə'nætəmik]　*adj.* 神经解剖学的
29. neurophysiological ['nju:rəu,fiziə'lɔdʒikəl]　*adj.* 神经生理学的
30. bioelectrical [,baiəui'lektrikəl]　*adj.* 生物电的
31. analgesia [,ænəl'dʒi:ziə]　*n.* 止痛
32. gastrointestinal [,gæstrəuin'testinəl]　*adj.* 胃肠的
33. immunological [,imjunə'lɔdʒikəl]　*adj.* 免疫学的
34. cardiovascular [,kɑ:diəu'væskjulə]　*adj.* 心血管的
35. hospice ['hɔspis]　*n.* 临终关怀医院
36. chronobiology [,krɔnɔbai'ɔlədʒi]　*n.* 生物钟学
37. energetics [,enə'dʒetiks]　*n.* 能量学
38. integrative ['intigreitiv]　*adj.* 综合的
39. complication [,kɔmpli'keiʃən]　*n.* 并发症
40. generic [dʒi'nerik]　*adj.* 普通的
41. synthesis ['sinθisis]　*n.* 综合
42. stringent ['strindʒənt]　*adj.* 追切的；严格的
43. pharmacological [,fɑ:məkə'lɔdʒikəl]　*adj.* 药理学的
44. intrinsic [in'trinsik]　*adj.* 本质的
45. holistic [həu'listik]　*adj.* 整体的

Useful Expressions

1. shed light on　阐明；使…清楚地显出
2. menstrual cramp　经期痉挛
3. carpal tunnel syndrome　腕管综合征
4. therapeutic intervention　介入疗法

5. molecular biology 分子生物学
6. herbal formula 药方
7. a wide array of 一系列
8. randomized controlled trial 随机对照实验
9. irritable bowel syndrome 过敏性大肠综合征
10. quasi-experiment 类实验
11. cohort study 群组研究
12. case-control study 病例对照研究
13. chronic illness 慢性疾病

Notes

1. 本文源自加州大学洛杉矶分校，作者许家杰，医学博士，加州大学洛杉矶分校医学院医学系教授，主任。

2. 中医药（traditional Chinese medicine），即传统医学（traditional medicine），也叫汉族医药，它是中华民族的宝贵财富，为中华民族的繁衍昌盛做出了巨大贡献。它包括中医和中药两个部分。简言之，中医是指在中医理论指导下，运用药物治疗疾病的学科；中药学则是研究中药的基本理论和临床应用的学科，是以中医理论指导的用药的学科。传统医学的治疗理念正逐渐为世界所接受，传统医药受到国际社会越来越多的关注，世界范围内对中医药的需求日益增长，这为中医药的发展提供了广阔的空间。

3. 草药（herbs），利用植物提取物制作而成，用来治疗疾病。利用草药减轻病痛是具有悠久历史的保健方式。

4. 中草药（包括中药和草药），在中国古籍中通称"本草"。我国最早的一部中药学专著是汉代的《神农本草经》，唐代由政府颁布的《新修本草》是世界上最早的药典。唐代孙思邈编著的《备急千金要方》和《千金翼方》集唐代以前诊治经验之大成，对后世医家影响极大。明代李时珍的《本草纲目》，总结了16世纪以前的药物经验，对后世药物学的发展做出了重大贡献。中药按加工工艺分为中成药和中药材。如今，随着对中药资源的开发和研究，许多民间药物也归入中药的范畴。所以中药是以中医理论为基础，用于防治疾病的植物、动物、矿物及其加工品，有着独特的理论体系和应用形式，充分反映了我国自然资源及历史、文化等方面的特点。

Unit 2　Traditional Chinese Medicine

Exercises

Part Ⅰ　Vocabulary and Structure

Section A　Match each word with its Chinese equivalent.

1. pharmacological   A. 种质
2. bioactive   B. 人参
3. antioxidant   C. 污染
4. purification   D. 品系
5. ginseng   E. 药理的
6. strain   F. 抗氧化剂
7. compatibility   G. 病理学
8. germplasm   H. 纯化
9. contamination   I. 配伍
10. pathology   J. 生物活性的

Section B　Fill in the blanks with the words or expressions given below. Change the form where necessary.

| perspective | efficacy | concurrent | encompass | adjunct |
| stringent | integrative | holistic | shed | derive |

1. However, what on earth has evoked four emotions of "Mona Lisa" _____?
2. A particular _____ is a particular way of thinking about something, especially one that is influenced by your beliefs or experiences.
3. The atmosphere _____ the Earth.
4. A (An) _____ professor is also a limited or part-time position to do research or teach classes.
5. Between technical and popular science, writing is what I call "_____ science", a process that blends data, theory and narrative.
6. He announced that there would be more _____ controls on the

possession of weapons.
7. The _____ of an alarm clock is explained as a very vigorous sensory input that triggers（引起）a large, synchronous（同时）assembly.
8. We have _____ a great deal of benefit from her advice.
9. They have one of the only _____ health programs in the country.
10. These discoveries may _____ light on the origins of the universe.

Part Ⅱ Translation

Section A Translate the following sentences into Chinese.

1. The current interest in traditional and complementary medicine in the United States is attracting attention in many parts of the community.
2. Many patients are using traditional and modern medical paradigms concurrently, creating a need for the appropriate and smooth merger of the two medicines.
3. Humans and animals have tested and used botanicals to relieve their suffering since ancient times.
4. Modern scientific investigations on plant-based medicine have been carried out in many parts of the world, including clinical trials of botanical combination products.
5. Harmonizing traditional medicine and modern medicine is more than utilizing modern research design or scientific technology to assess traditional medicine.

Section B Translate the following sentences into English.

1. 它用于人参中39种农药（pesticides）的测定。
2. 碱提取多糖（AEP）是从西洋参根中分离得到的。
3. 数十年来天然产物已成为抗菌（antibacterial）药物的丰富来源。
4. 多糖作为一种重要的生物活性成分，由于具有抗肿瘤（anti-tumor）、抗凝血（anti-clotting）、抗衰老（anti-aging）和免疫调节活性（immune modulating activity）等多种功能，引起了广泛关注。
5. 目前，根系化感物质（allelopathy of root）的研究成为了人参连作障碍（continuous cropping obstacle）成因的一个重要方面。

Text B

The Basic Characteristics of TCM Theoretical System

The theoretical system of traditional Chinese medicine (TCM) has evolved in the long course of clinical practice under the guidance of classic Chinese *materialism* (唯物主义) and *dialectics* (辩证法). ***It*** originates from practice and, in turn, guides the practice. This unique theoretical system is essentially characterized by the concept of *holism* (整体观) and treatment based on *syndrome differentiation* (辨证论治).

Concept of Holism

The concept of holism is a reflection of classic Chinese materialism and dialectics in TCM, emphasizing the ***integrity*** of the human body and the unity between the body and its external environments. The holism permeates through the physiology, pathology, *diagnostics* (诊断学), syndrome differentiation and *therapeutics* (治疗学).

The human body is regarded as an organic whole. Its constituent parts are inseparable in structure, interdependent in physiology, and mutually influential in pathology. The unity of the body is realized through the five zang-organs, with the assistance of the six fu-organs and the communication of the *meridian* (经络) system. Since the human body is an organic whole, treatment of a local disease has to take the whole body into consideration. For example, the heart opens into the tongue and is related with the small *intestine* (肠) internally and externally. So, oral erosion may be clinically treated by clearing away the fire from the heart or small intestine. There are a number of therapeutic principles in TCM developed under the guidance of the concept of organic whole, such as "drawing yin from yang and drawing yang from yin; treating the right for curing disease located in the left, treating the left for curing disease located in the right"; "needling the *acupoints* (穴位) on the lower part of the body for the treatment of the disease located in the upper part, and needling the acupoints on the upper part of the body for the treatment of the disease located in the lower part".

Man lives in the natural world and the natural world provides man with all the necessities indispensable to his existence. At the same time, the changes in nature directly or indirectly affect the human body. Take seasonal changes for example, usually spring is marked by warmth, summer by heat, late summer by dampness, autumn by dryness and winter by cold. Under the influence of such changes, the living things on the earth will also change to adapt to environmental variation, such as sprouting in spring, growing in summer, alternation in late summer, ripeness in autumn and storage in winter. The human body is no exception and it also makes corresponding changes to adapt to the changing seasons. For example, in spring and summer, yang qi goes outward and flourishes, qi and blood of the body tend to circulate superficially, consequently leading to more sweating and less urination. And during autumn and winter, yang qi goes inward and *astringes*（内敛）, qi and blood of body tend to flow internally, causing less sweating and more *urination*（排尿）. In this way the body keeps its balance of water metabolism and avoids over consumption of yang qi.

Treatment Based on Syndrome Differentiation

Treatment based on syndrome differentiation, another important feature of the theoretical system of TCM, is a basic principle in TCM for understanding and treating disease. Syndrome is the generalization of the progress of a disease at a certain stage. Since it involves the location, cause and nature of the disease, and the relation between *pathogenic*（致病的）factors and healthy qi, syndrome can comprehensively and accurately reveal the nature of the disease. Syndrome differentiation implies that the clinical data of a patient collected through the four examinations are analyzed and generalized so as to identify the pathological mechanism of the disease. Treatment means to select the corresponding therapy according to the result of syndrome differentiation. Taken as a whole, treatment based on syndrome differentiation is a process to understand and resolve a disease.

TCM emphasizes the differentiation of syndrome, because only when the syndrome is accurately differentiated can a correct treatment be made. Take common cold for example, its symptoms of fever, *aversion*（厌恶）to cold and pain in the head and body indicate that the disease is in the exterior. However, it is usually differentiated into two syndromes: common cold due to wind-cold and common cold due to wind-heat. For the treatment of the former syndrome in common cold, herbs *pungent* (有刺激性的) in taste and warm in nature are used; while for the treatment of the latter syndrome, herbs pungent in taste and cool in nature are used. So accurate differentiation of syndrome is the *prerequisite* (先决条件) for determination of a proper treatment. The core of treatment based on syndrome differentiation is to understand the relation between the nature and manifestation of a disease.

Since syndrome is the summarization of pathological changes of a disease at a certain stage of its course, there come the two important ideas. One is called "treating the same disease with different therapies", which means that one disease may manifest different syndromes at different stages or under different conditions, and thus needs to be treated by different therapies. Take measles for example, the treatment for it varies from stage to stage. The treatment focuses on promoting eruption at the early stage because it is not fully erupted, clearing away heat from the lung at the middle stage because lung heat is *exuberant* (旺盛的), and nourishing yin to clear away heat at the advanced stage because the heat still lingers and the lung and stomach yin is consumed. The other is known as "treating different diseases with the same therapy", which means that different diseases may present the same syndrome because they are of the same pathological mechanism, and thus may be treated with the same therapy. For example, *proctoptosis* (直肠脱垂) due to prolonged *dysentery* (脏毒) and *hysteroptosis* (子宫下垂) are two different diseases. But if they manifest the same syndrome of middle qi collapse, both of them can be treated by lifting middle qi.

It thus can be seen that the treatment of disease in TCM does not simply concentrate on the difference or similarity of diseases, but on the difference or similarity of *pathogenesis* (发病机理). This indicates that diseases with the same pathogenesis can be treated with the same therapeutic methods, while diseases with different pathogenesis have to be treated with different therapeutic methods. Such a therapeutic principle is the *gist* (本质) of treatment based on syndrome differentiation.

(1,083 words)

Comprehension of the Text

Choose the best answer to each of the following questions.

1. During the development of TCM, _____ are the guidance of its theory.

    A. Classic Chinese materialism and dialectics

    B. Traditional theory and practice

    C. Yin and yang theories

    D. Holism and syndrome differentiation

2. What does the italicized and boldfaced word "It" in paragraph 1 refer to?

    A. Classic Chinese materialism and dialectics.

    B. The theoretical system of TCM.

    C. The long course of clinical practice.

    D. Five elements theory.

3. Which of the following words has the similar meaning with the italicized and boldfaced word "integrity" in paragraph 2 ?

    A. Individuality.　　　　　B. Comprehensibility.

    C. Unity.　　　　　　　　D. Complexity.

4. Which of the following is not the constituent part of the human body?

    A. Five zang-organs.　　　B. Six fu-organs.

    C. Meridian system.　　　D. Spirit.

5. With the concept of holism, why can oral erosion be treated by curing the disease from the heart or small intestine?

    A. Because the heart opens into the tongue.

    B. Because we should draw yin from yang.

    C. Because we can treat the lower part of body for curing disease located in the upper part.

    D. Because clearing away the fire from the heart is important.

6. With the changing seasons, _____ should be kept well during autumn and winter.

    A. blood　　　　　　　　B. sweating

    C. yang qi　　　　　　　D. water

7. What is the fundamental principle in treating disease?

    A. Cause and nature of the disease.

B. Syndrome differentiation.

   C. The concept of holism.

   D. Treating different diseases with the same therapies.

8. Why is differentiation of syndrome so important in TCM?

   A. Because in order to make a correct treatment, syndrome must be distinguished.

   B. Because the human body is considered as an organic whole.

   C. Because the changes in nature affect the human body directly or indirectly.

   D. Because syndrome is the summarization of a disease progress at a certain stage.

9. What kind of herbs can be used to treat common cold due to wind-heat?

   A. Herbs pungent in taste and warm in nature.

   B. Herbs pungent in taste and cool in nature.

   C. Both A and B.

   D. None of the above.

10. Treating measles with different therapeutic measures at different stages is an example of _____ .

    A. treating different diseases with the same therapy

    B. treating different diseases with different therapies

    C. treating the same disease with the same therapy

    D. treating the same disease with different therapies

科技文体翻译技巧（二）

翻译的转换

为了达到更好的翻译效果，我们常使用一种翻译手段，即翻译中的转换。它包括引申、转换、正反表达等。英汉两种语言在文化背景、修辞手法、句子结构和词汇使用上有很多的差异，因此翻译中的转换是不可避免的。

一、引申法

词义的引申是指词义的转换。英汉两种语言在词汇使用上有很多不同之处,翻译时在词典里常找不到完全对等的词语。如果望文生"译",译文必然生硬难懂,不但不能准确表达原意,甚至可能造成很大的误解。因此在翻译时,必须从上下文中的逻辑关系和词的基本词义出发,进一步引申其深层含义,才能选择出恰当的词语来表达。词义的引申不只是要从上下文的逻辑关系出发,还要考虑到选词搭配、句子结构、修辞习惯、文化背景和原作者背景等因素,然后再选择恰当词语表达。

【例】1. This could give more harvest of fruit and increase soil fertility.

这既能增加果实产量,又能培养地力。

2. As the straw of bean is used as manure, the production of early corn can be increased by 14.4%.

将豆秆还田,对早玉米增产效果明显,可增产 14.4%。

二、转换法

英汉两种语言在词汇搭配、句子结构和修辞手法上有一定的差异,互译时不一定能完全保持源语与目的语中词性的一致。因此为了适应译文语言的需要,在翻译中我们往往要运用词性转换以及语态转换手段来保持译文语言的规范性。

(一)词性转换

词性转换通常与句子成分或语态的转换有关,句子结构的变化也会造成句子成分的转换和词类的转换。例如名词可转换成动词或形容词,动词可以转换成名词,形容词或介词,形容词和副词可以互相转换。在科技文献的翻译中,最重要的是内容和形式上的协调一致。

【例】1. The operation of a car needs some knowledge of its performance.

操作汽车需要懂得它的一些性能。

2. At constant temperature, the pressure of a gas is inversely proportional to its volume.

温度不变,则气体压力与其体积成反比。

(二)语态的转化

语态的转化是一种常见的翻译手段。主动句和被动句可在一定情况下相互转换。因为英语中的科技语体的特点,在科技文献的英译中,将主动句译为被动句是较为常见的。科技文献的作者和读者主要着眼于逻辑推理的过程和演绎论证的结果,叙事说理也着重于事物间相互关系及演变过程,并不重视动作的发出者,即施动者。因此在汉语科技文体中广泛使用省略主语的动宾句或自然被动句,而英语中则广泛使用被动句。

【例】1. It is earnestly hoped that there will be more researchers doing some more systematic and intensive studies in the fields of domestic animal ecology in our country.

希望能有更多科研人员把我国家畜生态学研究系统而深入地开展起来。

2. These computers should be tested under normal working conditions.

这些计算机应在正常运转状态下测试。

汉语中的自然被动句表面上看是主动语气，但实际上是被动句。例如"实验做完了"，此句译成英语是通常译为被动句。类似的汉语表达还有"分为""变为""位于""成为""予以""与…相…""出现""旨在"等，英译时一般都译为被动语态。例如"它们位于表层土以下"译为"They are located under top soil"；"AC 与 BD 相连"译为"AC is linked to BD"。一些句子中的主语可以译为英语句子中的被动语态谓语动词。

【例】1. The electric current is defined as a stream of electrons flowing through a conductor.

电流的定义是流经一个导体的电子流。

2. 100 feet was read from the altimeter.

高度表的读数是 100 ft（英尺）。

另外，汉语中有些谓语动词可以译为英语"主语＋被动语态谓语动词"的结构。

【例】Full advantage was taken of these opportunities by the participants, as a result, many lively academic discussions ensured.

与会者充分利用这个机会，进行了许多生动活泼的学术讨论。

三、正反表达法

正反表达法是指在汉译英时，为了适应英语语言表达的需要，必须使用与原文意思相反的词汇或句式，以准确表达原文的整体含义。

（一）反译法

反译法有两方面的含义，一是句子结构上，一是逻辑上。在句子结构上，否定句和肯定句相互转换。

【例】1. Gold does not melt until heated to a definite temperature.

黄金加热到一定温度才会熔化。

2. Egg is rich in protein, while rice and bread are not.

鸡蛋含蛋白质丰富，而米饭和面包含蛋白质较少。

在逻辑上，正面表述与反面表述相互转换。

【例】1. These experimental values agreed with the theoretical values within the accuracy of ±0.05%.
这些实验的数值与理论值相符，误差在±0.05%范围内。
2. There is no law that has no exceptions.
凡是定律都有例外。
3. Researchers did a lot research on the culture of endosperm.
科研人员们在植物胚乳培养方面做了不少研究工作。

（二）否定形式的转换

汉语中的否定词往往置于动词前来否定谓语，而英语的否定词可以放在动词前，还可以放在名词或其他词类之前来否定主语、宾语、状语等。英语中构成否定的方式有多种，包括句法手段、构词手段、词汇手段等。句法手段构成否定是通过加入否定词来实现；构词手段指通过给词语添加前缀或后缀实现；词汇手段是指使用某些具有否定意思的词。英语中的否定句可分为完全否定、部分否定、双重否定等。而否定句的使用没有明显界限。否定形式的选择主要取决于否定语气的轻重和句子的结构。

【例】1. One often does not believe air to have weight.
人们通常认为空气没有质量。
2. There is little oil left in the tank.
油箱里几乎没有什么油。

英汉否定方式不同，翻译时常会转换否定成分。

1. 主语否定转换为谓语否定

【例】No energy can be created and none destroyed.
能量既不能创造也不能消灭。

2. 宾语否定转换为谓语否定

【例】Scientists know of no effective way to store solar energy.
科学家们还不知道储藏太阳能的有效方法。

3. 谓语否定转换为补语否定

【例】The machine does not work properly.
这台机器运转得不正常。

4. 谓语否定转换为状语否定

【例】The machine did not stop running because the fuel was finished.
这台机器并非因为燃料用完才停止运转的。

科技英语摘要写作（二）

指示型摘要

科技文体中指示型摘要也被称为概括型摘要或描述型摘要。指示型摘要多用于理论性较强的论文。它定性地指出论文所探讨的对象、目的、角度、方法、主要结论等，并不定量地指出论文中所报道的具体内容。它仅能使读者决定是否需要阅读全文，而不能从中直接获取定量信息。这类摘要篇幅较短，一般只有一个自然段。它叙述了论文的主要内容，但不说明具体方法、结果、结论或建议。仅指出文章的综合内容，一般适用于综述性文章、图书介绍以及编辑加工过的专著等，最常见的有某技术在某时期的综合发展情况或某技术在目前的发展水平及未来展望等。以下列一个范文。

Abstract：The Body View of TCM is the comprehensive understanding about the life. Through the study on the body view of TCM, not only the knowledge of the structure and function of the human body, but also the penetrating of the Traditional Chinese Culture can be found out. Therefore, it is a very important field in the correlation research between the TCM and the Traditional Chinese Culture. This article analyzed the relationally social and cultural backgrounds in the process of the building and developing of the TCM theories, and elaborated the influence of the Traditional Chinese Culture in the TCM through the study on the Kidney and the Life Gate in the view of the ideological history. The author focuses on the content and features of the body view of TCM, as well as its construction methods and its evolution at different times.

摘要译写示例

示例一

【摘要中文原文】

摘要：以人参根愈伤组织为受体，分析了无土栽培条件下收集的人参原生根系分泌物的自毒作用。结果表明：（1）人参根系分泌物可抑制愈伤组织的生长，且随处理浓度增加，抑制活性增强，200 mg/L 时，生物量降低 56.1%，与对照差异显著。（2）影响了受体抗逆酶活性强弱及变化趋势，短时间或低浓度处理下，酶活性增强，其中 SOD 酶活性增加幅度最大，其次为 POD 和 CAT，而长时间或中高浓度处理后，酶活性均显著降低；同时 3 种酶原有的协调作用关系及平衡被破坏。结论：人参原生根系分泌物具有化感自毒活性，对受体抗逆酶系统的显著影响是人参根系分泌物化感活性的重要表现。

【摘要原版译文】

Abstract：To study autointoxication of primary root exudates which was collected by soilless cultivation with the natural equivalent of Ginseng root callus as receptor. The result to：(1) Ginseng root exudates showed inhibitory effects on the growth of callus and the higher the concentration, the stronger inhibition. Biomass was 56.1% lower in 200 mg/L which had significant difference with control group. (2) Ginseng root exudates could affect the strength and trend of stress-resistant enzyme：the activity of stress-resistant enzyme could increase at short time or low concentration. And the activity of SOD increased most which was followed by POD and CAT. The activity of stress-resistant enzyme depressed significantly at long time or medium and high concentration； and coordinating functions of stress-resistant enzyme were destroyed. Conclusion：ginseng primary root exudates had inhibitory effects on activity of stress-resistant enzyme which is the important expression for autotoxicity of ginseng primary root exudates.

【摘要修改译文】

Abstract：The autointoxication of primary root exudates collected by soilless cultivation was analyzed with the natural equivalent of ginseng root callus as receptor. The results showed that (1) Ginseng root exudates showed inhibitory effects on the growth of callus and the higher the concentration, the stronger the inhibition. Biomass decreased by 56.1% at 200 mg/L, which was significantly different from the control group. (2) Ginseng root exudates could affect the strength and variation trend of stress-resistant enzyme：the activity of stress-resistant enzyme could increase under short time or low concentration treatments. And the activity of SOD increased by the largest amount, followed by POD and CAT. The activity of stress-resistant enzyme depressed significantly under long time or medium and high concentration treatments； coordinating functions of stress-resistant enzyme were destroyed. In conclusion, ginseng primary root exudates had inhibitory effects on the activity of stress-resistant enzyme, which is the important expression for autotoxicity of ginseng primary root exudates.

【主要修改意见】

1. 避免望文生"译"。如第一句中"To study autointoxication of primary root..."应以名词性短语为主语。
2. 语句力求简洁。善于使用后置定语以减少句中的定语从句,如"which was collected by..."可以译为"collected by"。
3. 固定表达要规范。如"结果表明"应译为"The results showed that";分号后可省略"and",以达到简洁表述的目的。

示例二

【摘要中文原文】

摘要:以人参6种病原真菌及其6种拮抗细菌和4种拮抗放线菌为受体,研究了新林土、三年生人参根际土壤和撂荒十年老参地土甲醇提取物对其生长的影响。结果表明:(1)人参根际土壤提取物对病原真菌生长的影响存在抑制和促进2种类型,对拮抗细菌、拮抗放线菌的影响表现为抑制;(2)对受体真菌、细菌、放线菌的影响存在明显的种间差异;(3)综合比较抑制作用强弱顺序为:拮抗细菌>拮抗放线菌>病原真菌。总体看来,新林土本身存在着影响受体微生物生长的物质组分,栽培人参后,该组分活性增强,表明人参根际土壤中物质和组成的变化与人参病原真菌、拮抗细菌、拮抗放线菌的生长具有一定的相关性。

【摘要原版译文】

Abstract: The influence on growth of six kinds of ginseng pathogenic fungi, six kinds of antagonistic bacterias and four kinds of antagonistic actinomycetes was studied by the methanol extracts of new forest soil, the ginseng rhizosphere soil for three years of continuous cropping and the ginseng-grown land soil for ten years of rest. The results showed that: (1) The methanol extracts of ginseng rhizosphere soil presents inhibitory effect, or promoting effect occasionally on ginseng pathogenic fungi, while only presenting inhibitory effect on antagonistic bacteria and actinomycetes. (2) The inhibitory and promoting effects were obviously different between test microbial species. (3) The inhibitory effects fall into the following sequence: antagonistic bacterias >

antagonistic actinomycetes ＞ pathogenic fungi. Overall, some chemical components are present in new forest soil affecting the growth of receptors. Once ginsengs being planted, the activity of the chemicals in soil increases. The growth of ginseng pathogenic fungi, antagonistic bacterias and actinomycetes is evidently affected by change of some chemical substances and their composition in ginseng rhizosphere soil.

【摘要修改译文】

Abstract: The effects of the methanol extracts of new forest soil, three-year ginseng rhizosphere soil and ginseng-grown land soil resting for ten years on growth of six kinds of ginseng pathogenic fungi, six kinds of antagonistic bacteria and four kinds of antagonistic actinomycetes were studied. The results showed that: (1) The methanol extracts of ginseng rhizosphere soil present inhibitory effect, or promoting effect occasionally on ginseng pathogenic fungi, while only presenting inhibitory effect on antagonistic bacteria and actinomycetes. (2) The inhibitory and promoting effects are obviously different between test microbial species. (3) The inhibitory effects fall into the following sequence: antagonistic bacteria＞antagonistic actinomycetes＞pathogenic fungi. Overall, some chemical components are present in new forest soil affecting the growth of receptors. The activity of the chemicals in soil increases after ginsengs are planted. The growth of ginseng pathogenic fungi, antagonistic bacteria and actinomycetes is evidently affected by changes of some chemical substances and their composition in ginseng rhizosphere soil.

【主要修改意见】

1. 科技文体中"影响"一词译为"effect"较为准确。
2. 译文力求简洁。如"总体看来"可直接译为"overall"。
3. 长句需要按照意群拆分。如"总体看来,新林土本身存在着影响受体微生物生长的物质组分,栽培人参后,该组分活性增强,表明人参根际土壤中物质和组成的变化与人参病原真菌、拮抗细菌、拮抗放线菌的生长具有一定的相关性",可按照意群译为3个短句。
4. 注意英语中名词的单复数变化。如"bacteria"即为复数形式。

Unit 3 Education

Text A

The Impact of COVID-19 Pandemic on Education: Social Exclusion and Dropping out of School

Introduction

An important historical time is being experienced by people worldwide, imposing a sudden quarantine and leaving them with no time to either realize what that meant or what its consequences would be. What was familiar was quickly abandoned while the need for immediate adaptation to a new way of living that of social isolation and physical distancing emerged. Complexity and uncertainty are the "new normal" and could even be for quite a period of time.

COVID-19 pandemic has not only been a major health crisis but an educational one as well. The specific crisis has posed on policy makers the dilemma of either shutting down educational institutions (schools—both state and private—and universities) or keeping them open, thus either saving lives by eliminating social contacts or maintaining every state's economy. The decision that most countries worldwide have made has been to proceed with the lockdown of all educational institutions. Consequently, the physical doors of schools globally were closed and both teachers and students were abruptly forced to go into remote teaching and learning mode.

Such a decision though beneficial for the interception of each pandemic wave carries the risk of being detrimental for the smooth learning process and the overall development of students. The implementation of synchronous, distant learning has been adopted by the majority of the affected countries, resulting in bringing out one of the most serious problems tormenting societies for ages, that is, the problem of socially and educationally disadvantaged children. It has been proved that students coming from families with a low socio-economic status are likely to be severely affected by such conditions. The deprivation of appropriate technological infrastructure, the absence of internet access, the lack of basic digital skills, or even the shortage of a quiet studying space are all constraining factors of those students' equal participation in distant learning. The financial pressure put on their families—more powerful and painful for the financially deprived families—is likely to inflict a major blow on them, resulting in their dropping out of school.

COVID-19 Pandemic and Dropping out of School

A few months ago, an unpredictable situation COVID-19 virus pandemic reshaped the way we live, think and communicate. With coronavirus cases growing rapidly in the affected countries, we had no other option but to adjust our way of thinking and living accordingly, trying to come to terms with a new status of living. Education, of course, could not be the exception and radical changes and adaptations had to be made asap. Concerns have been raised not only by education stakeholders but also by health experts, teachers and parents while the debate about the reopening of schools has been ongoing. Teachers and parents worry about learning loss that will not easily be made up for even if schools quickly return to their prior performance levels, and health issues as education leaders are struggling to find ways of keeping the learning process going. New laws are rapidly being enforced so as distant learning is to replace the teaching and learning process in classrooms, as the social importance of schools has been recognized by all, even health experts' associations. Consequently, the political decision of closing down all educational institutions until further notice has led to growth and increased use of synchronous and asynchronous methods of teaching and learning. Students are called upon to get used to a new teaching and learning routine, being taught from the comfort of their homes and trying to get in grips with new digital tools of distant learning, such as Webex, Skype, Zoom, Messenger, etc.

However, the quality of education students receive is mainly dependent on their access to these digital learning resources, unveiling a hard reality for a great number of students and their families. The fact that only 60% of the global population are in the position to use digital tools cannot be ignored. Thus, societal issues of inequality and uninhibited access to the benefit of education have been manifested and collectively witnessed, even if these issues have been pinpointed in several debates on exclusion in education for years.

Similarly, although levels of access to information technology have boomed over the past 10 years, there is considerable evidence of a so-called digital divide, namely, a growing disparity between those individuals and communities that have and those that do not have easy access to new information technologies. Moreover, there are many more obstacles to be surpassed apart from just a student's capability of using technology. Some of these may include families that only have one internet accessible device at home and more than one child who need to use it simultaneously or even the space to offer them as a quiet learning environment or even underprivileged families that do not have such devices available to be used for e-learning. Poor housing, in particular overcrowding, access to basic amenities, and temporary accommodation are also associated with lower educational attainment. The effort made at home by students to retain their learning pace given the lack of basic technological equipment to connect and participate in distant learning, the weakening of their basic learning skills due to their long absence from school and its corresponding lack of practice, the difficulties they face in trying to carry out the educational activities assigned to them remotely, but without having the appropriate help from their family environment in order to respond, the lack of supportive motivation and, ultimately, the limitation of their educational expectations due to the uncertainty that is formed in their learning environment, all comprise factors that create an explosive mixture in their effort to maintain their student identity.

An important parameter of the issue of the children's school exclusion due to COVID-19 pandemic and its consequent digital divide created is the children who have been experiencing social and educational exclusion for years, that is, the children with disabilities and/or special needs, of single-parent families, of various races and ethnicities as well as the Gypsy/Roma children. Official statistics show that these children are over three times more likely to be excluded from school than the school population as a whole. A UK-wide literature review by Wilkin, Derrington, White, Martin, Foster, Kinder, & Rutt supports that exclusion from school is similar to overall levels of social exclusion for these groups. It is also claimed that excluding children who belong in the aforementioned groups can actually trigger further exclusion, as one of the catalysts underpinning their exclusion is their relatively low school attendance.

The current situation of closing down schools for long periods of time and its subsequent swift to distant learning have widened the gap of inequity for these underprivileged students. Low or no attendance can affect students' ability to keep up with schoolwork and makes them more likely to fall behind academically. In other words, punishing these students by school exclusion directly increases their risk of further exclusion, due to poor attendance and falling behind educationally, which can aggravate a tendency for these groups to be subsequently excluded at a society level, which is already a risk. Such conditions adversely affect upon a child's health, development and access to friends and social networks, which are likely to affect school attendance and performance. According to latest study reports, approximately 1.5 billion students are out of school across 186 countries in the world in the era of COVID-19 pandemic. If students do fall behind, they have few opportunities to make up lost ground in the current education system. Such students tend to fall further behind their peers and may give up altogether the attempt to catch up. Moreover, the longer-term consequences of school exclusion are often profound. Research has shown that students who are excluded from school in the final two years of compulsory education are two and a half times as likely not to participate in education, training or employment between the ages of 16 and 18 than those not excluded.

Conclusion

In our times, the unhindered access to the educational system is considered, more than ever, an inalienable social right and the abolition of educational exclusion at all levels of education is the beginning of every educational policy. However, unexpected, severe phenomena, as COVID-19 pandemic has been, have threatened the unimpeded school attendance by millions of students globally, widening the educational and social disparity among them, rendering the underprivileged children more at risk of abandoning school prematurely. The purpose of the present study was to put emphasis on the impact that this pandemic has on children who suffer educational and social exclusion, forcing them to drop out of the learning process as this has taken a different shape under the specific circumstances. Our hope is that this study, which sees the light of day in a difficult time for humanity, will contribute to raising the awareness of all stakeholders of all children's right to equally participate in education.

(1,591 words)

New Words

1. pandemic [pæn'demik] *n.* 流行病
2. impose [im'pəuz] *vt.* 强加
3. quarantine ['kwɔrəntiːn] *n.* 隔离
4. dilemma [di'lemə] *n.* 困境
5. eliminate [i'limineit] *vt.* 消除
6. lockdown ['lɔkdaun] *n.* 封锁
7. interception [ˌintə'sepʃn] *n.* 拦截
8. detrimental [ˌdetri'mentl] *adj.* 不利的
9. synchronous ['siŋkrənəs] *adj.* 同步的
10. torment ['tɔːment] *vt.* 困扰, 折磨
11. deprivation [ˌdepri'veiʃn] *n.* 缺乏
12. infrastructure ['infrəstrʌktʃə(r)] *n.* 基础设施
13. constrain [kən'strien] *vt.* 限制
14. inflict [in'flikt] *vt.* 使遭受打击
15. coronavirus [kəˌrəunə'vaiərəs] *n.* 冠状病毒
16. stakeholder ['steikhəuldə] *n.* 参与人
17. asynchronous [ei'siŋkrənəs] *adj.* 不同时存在（或发生）的
18. unveil [ˌʌn'veil] *vt.* 展示
19. uninhibited [ˌʌnin'hibitid] *adj.* 不受限制的
20. manifest ['mænifest] *vt.* 显示, 表明
21. pinpoint ['pinpɔint] *vt.* 精确找到
22. boom [buːm] *vi.* 迅速发展
23. considerable [kən'sidərəbl] *adj.* 相当大的
24. disparity [di'spærəti] *n.* 差距
25. underprivileged [ˌʌndə'privəlidʒd] *adj.* 贫困的
26. amenity [ə'miːnəti] *n.* 便利设施
27. comprise [kəm'praiz] *vt.* 构成

28. ethnicity [eθ'nisəti] n. 种族特点
29. aforementioned [ə,fɔ:'menʃənd] adj. 前面提及的
30. catalyst ['kætəlist] n. 催化剂
31. underpin [,ʌndə'pin] vt. 巩固支持
32. academically [,ækə'demikli] adv. 学术上
33. aggravate ['ægrəveit] vt. 加重
34. unhindered [ʌn'hindəd] adj. 畅通无阻的
35. inalienable [in'eiliənəbl] adj. 不可剥夺的
36. unimpeded [,ʌnim'pi:did] adj. 畅通无阻的
37. prematurely ['premətʃə(r)li] adv. 过早地

Useful Expressions

1. COVID-19 pandemic 新型冠状病毒肺炎大流行
2. social exclusion 社会排斥
3. shut down 关闭
4. get in grips with 掌握
5. have access to 可以利用
6. temporary accommodation 临时住所
7. keep up with 跟上

Notes

1. 本文选自 https://www.scirp.org/journal/paperinformation.aspx?paperid=107598，作者：Olympia Tsolou, Thomas Babalis, Konstantina Tsoli.

2. COVID-19：新型冠状病毒肺炎，症状为发热、干咳、呼吸急促及呼吸困难。重症者病情进展迅速，数日内即可出现急性呼吸衰竭、肾衰竭而危及生命。

3. 在线教育（online education）：在线教育或称远程教育、在线学习，一般指一种基于网络的学习行为。在线教育顾名思义，是以网络为介质的教学方法。在线教育跨越时空和人力物力限制，具有资源利用最大化、学习行为自主化、师生交流与学生自学等学习形式交互化、教学管理自动化等诸多优势与特点，是一种教育的巨大变革。

Exercises

Part Ⅰ Vocabulary and Structure

Section A Match each word with its Chinese equivalent.

1. pandemic
2. academically
3. asynchronous
4. eliminate
5. boom
6. inflict
7. constrain
8. stakeholder
9. unveil
10. underprivileged

A. 不同时存在(或发生)的
B. 消除
C. 使遭受打击
D. 限制
E. 参与人
F. 学术上
G. 展示
H. 贫困的
I. 流行病
J. 迅速发展

Section B Fill in the blanks with the words or expressions given below. Change the form where necessary.

| considerable | dilemma | deprivation | aggravate | unimpeded |
| interception | synchronous | comprise | prematurely | boom |

1. A _____ is a difficult situation in which you have to choose between two or more.
2. Information sent through the internet is vulnerable (易受攻击的) to third-party _____ .
3. Millions more suffer from serious sleep _____ caused by long work hours.
4. A _____ programming model is much simpler than an asynchronous one.
5. Doing it properly makes _____ demands on our time.
6. The committee is _____ of representatives from both the public and private sectors.
7. If the reports are well founded, the incident could seriously _____ relations between the two nations.

8. The heavy rain driving against the windows made the room _____ dark.
9. U.N. aid convoys (联合国救援车队) have _____ access to the city.
10. If the economy or a business is _____, the number of things being bought or sold is increasing.

Part Ⅱ   Translation

Section A   Translate the following sentences into Chinese.

1. An important historical time is being experienced by people worldwide, imposing a sudden quarantine and leaving them with no time to either realize what that meant or what its consequences would be.
2. COVID-19 pandemic has not only been a major health crisis but an educational one as well.
3. Such a decision though beneficial for the interception of each pandemic wave carries the risk of being detrimental for the smooth learning process and the overall development of students.
4. A few months ago, an unpredictable situation COVID-19 virus pandemic reshaped the way we live, think and communicate.
5. Research has shown that students who are excluded from school in the final two years of compulsory education are two and a half times as likely not to participate in education, training or employment between the ages of 16 and 18 than those not excluded.

Section B   Translate the following sentences into English.

1. 一些倾向建构主义学习观念（constructivist learning）的教师认识到了使用适当方法评价学生在线课程表现的重要性。
2. 20世纪末期，随着计算机技术的日益成熟及其在教育方面的广泛应用，多媒体技术在我国高等教育教学中出现，并迅速改变传统的教学方式。
3. 在教育信息化的新形势下，强调信息技术应用的翻转课堂（flipped classroom）被迅速引用到双语（bilingual）教学新模式中。
4. 交互性教学是对传统教学模式的突破，强调学生在学习过程中的自我探索和互动合作。
5. 与任何教学媒介（instructional medium）一样，远程教育有其优点和缺点，而这些必须被纳入远程教育计划的总体规划之中。

Text B

Artificial Intelligence in Education

Developments in science and technology in recent years have deeply affected societies. One of these developments is artificial intelligence, which is successfully used in many industries today. Artificial intelligence has become increasingly important and inspired many *interdisciplinary* (跨学科的) studies recently, and it is one of the up-to-date study fields of today. Artificial intelligence, which first emerged based on concepts such as the ability of computers or robots to think and feel, has been a field of study that has made incredible progress in the last 50 years. Today, artificial intelligence studies have become one of the areas that countries attach the most importance in their investments. It attracts more and more attention every day with the contribution of successful projects in recent years. The concept of artificial intelligence, which was first voiced by John McCarthy at the Dortmund conference in 1956, has started to be accepted as the technology of the future globally and at country level recently. Artificial intelligence, one of the important research areas in the field of computer engineering, can be considered as the driving force of technology since the first half of this century.

The term artificial intelligence was first defined by John McCarthy as intelligent machines, and particularly the science and engineering of making intelligent computer programs. In general terms, artificial intelligence is the ability of computers to perform higher *cognitive* (认知的) functions peculiar to human intelligence, such as *perception* (感知), decision-making, problem-solving, generalizing, gaining experience and acting accordingly. Artificial intelligence can be named as systems that imitate the human brain in order to fulfill the specified tasks and that can improve itself *recursively* (递归地) thanks to the experience gained as a result of the task. With regards to education, artificial intelligence is defined as information processing systems that can be involved in processes carried out by humans such as learning, adaptation, synthesis, self-correction, and the use of data for complex processing tasks. Even if different definitions are made about this concept, the common point is that artificial intelligence represents systems that can make human-like decisions by *emulating* （仿效） human thinking processes. The aim of artificial intelligence is to imitate human intelligence through computers, in this sense, to give computers the ability to learn. However, most of the artificial intelligence studies model the brain and intelligence of the humans, who are accepted as the most intelligent creatures known in the world.

With an increasing *momentum* （势头） since the 1960s, artificial intelligence studies have begun focusing more on the realistic and achievable goal of developing *algorithms* （算法）, programs, and systems that can model the problem-solving skill of the human brain rather than producing machines that think like humans. This has led to the development of many different artificial intelligence technologies such as artificial neural networks, expert systems, agents, fuzzy logic, natural language processing, Bayesian networks, genetic algorithms, deep learning, machine learning, speech recognition, computer vision. Although most of them are at the level of laboratory studies, it is mentioned that there are more than 60 artificial intelligence technologies today. However, there has been no *consensus* （共识） so far on establishing the standard approach to find out which technique has the most appropriate artificial intelligence learning theory for a particular learning environment. Moreover, scientists have not yet developed a piece of software to make it easier to determine learning style using the students' learning behavior as a model.

With the investments and technological developments made over the years, artificial intelligence applications have started to be used in many areas and have provided great convenience in these areas. Artificial intelligence applications are used in many fields such as industry, energy, health, banking, transportation, engineering, defense, and security. By using artificial intelligence in these areas, time and cost savings become possible, and this contributes significantly to production by minimizing error rates. It can be said that artificial intelligence has started to show itself in the education and can provide important developments in this field. Artificial intelligence applications in education are widely used by students and educators today, including various tools and applications such as smart lesson systems, teaching robots and adaptive learning systems. According to the 2018 Horizon report, artificial intelligence is one of the important developments in the field of educational technologies. In particular, over the past two decades, the increasing trend of big data and machine learning has significantly contributed to the educational development of artificial intelligence. The incorporation of artificial intelligence technology in education has broadened the scope of education, which was previously a human-oriented system. It is worth mentioning that studies on the use of artificial intelligence in education have increased in numbers. However, more research is necessary to ensure successful practice of artificial intelligence applications in the field of education.

It is believed in the literature that artificial intelligence applications add a different dimension to education and have many benefits. First, it can be said that artificial intelligence enables individualization of the learning process. Moreover, with artificial intelligence, students with learning difficulties can be identified at an early stage, and special solutions can be produced for them. Thus, effective teaching practices can be realized for students with special needs. Among the *prominent* （突出的） benefits of artificial intelligence applications is the fact that they offer excellent observations and *inferences* （推断） very quickly and at minimum cost. Besides, artificial intelligence is also used in the evaluation of student participation and academic *integrity* （诚实）. Furthermore, it prevents loss of time for teachers, and provides the opportunity to easily collect and store student data. On the other hand, artificial intelligence is used in the process of monitoring students' performance. Additionally, artificial intelligence is also used to provide useful feedback to students and teachers and to assist students in improving their academic writing skills. Artificial intelligence can raise awareness among school administration and teachers by giving early warning about unwanted student behavior and performance. Using artificial intelligence, teachers can analyze students in a classroom and understand who a slow learner is. In some areas where the student is weak or unsuccessful, appropriate steps can be taken to support learning with artificial intelligence analysis. Therefore, these potentials of artificial intelligence in the field of education have led to investments in many countries to further develop artificial intelligence and to increase the number of studies in recent years.

(1,178 words)

Comprehension of the Text

Choose the best answer to each of the following questions.

1. Which of the following expressions about artificial intelligence is INCORRECT?

    A. Artificial intelligence is successfully used in many industries today.

    B. Artificial intelligence has become increasingly important and inspired many interdisciplinary studies recently.

    C. Artificial intelligence is one of the updated study fields.

    D. Artificial intelligence has been a field of study that has made incredible progress in the last two decades.

2. In which year did John McCarthy first voice the concept of artificial intelligence?

    A. In 1954.　　B. In 1955.　　C. In 1956.　　D. In 1957.

3. With regards to education, artificial intelligence can do the following EXCEPT _____.

    A. learning　　　　　　　　　B. adaptation

    C. decision-making　　　　　D. self-correction

4. Since the 1960s, what have artificial intelligence studies begun focusing more on?

    A. Modeling the brain and intelligence of the animals.

    B. Developing algorithms, programs, and systems that can model the problem-solving skill of the human brain.

    C. Producing machines that think like humans.

    D. Developing many different modes of mentalities.

5. According to the passage, what consensus has not been achieved on so far?

    A. Establishing the standard to measure students' learning activities.

    B. Developing a software to make it easier to determine learning style using the students' learning behavior as a model.

    C. Finding out which technique is the best to combine artificial intelligence with education.

    D. Figuring out the appropriate artificial intelligence learning theory for a particular teaching environment.

6. Which of the following is NOT a benefit from the use of artificial intelligence in many fields?

Unit 3　Education

   A. Time-saving becomes possible.
   B. Cost-saving becomes impossible.
   C. It contributes significantly to production by minimizing error rates.
   D. It provides important developments in education.
7. Artificial intelligence contributes to education in many ways except that _____ .
   A. various tools and applications are widely used by students and educators today
   B. smart lesson systems, teaching robots and adaptive learning systems are used in some classrooms
   C. the increasing trend of big data and machine learning has significantly contributed to the educational development of artificial intelligence
   D. the incorporation of artificial intelligence technology in education has narrowed the scope of education
8. Within the field of education, what can artificial intelligence do to enable students?
   A. Artificial intelligence enables generalization of the learning process.
   B. Students with learning difficulties are identified at the final stage and offered special solutions.
   C. Effective teaching practices can be realized for students with special needs.
   D. Observations and inferences are done slowly but at minimum cost.
9. Which of the following statements is INCORRECT?
   A. Artificial intelligence is used to evaluate student participation and academic integrity.
   B. Artificial intelligence makes it easy for teachers to collect and store student data.
   C. Artificial intelligence is used in the process of monitoring students' performance.
   D. Artificial intelligence is used by teachers to provide harsh criticism to students.
10. In what kind of area appropriate steps can be taken to support learning with artificial intelligence analysis?
    A. In the area where there is no modern information technology.

B. In the area where the curriculum is designed unsystematically.

C. In the area where the student has difficulty in learning.

D. In the area where the teacher adopts traditional teaching methods.

科技文体翻译技巧（三）

从句的译法

科技英语中的复合句一般分为三大类型：名词性从句、形容词性从句和副词性从句。

一、名词性从句

在整个复合句中，起名词作用，充当主语、宾语、表语和同位语等的各种从句，统称为名词性从句。名词性从句主要有以下几种。

（一）主语从句

主语从句通常由以下两种方式构成。

1. 从句（由关联词或从属连词位于句首引导的从句）＋谓语动词＋其他成分

由于在主从复合句中起到主语的作用，因此这些句子在翻译时一般放在句首。关联词通常有 where、whatever、how、why、whoever、who、whenever、what、which、wherever 等，而从属连词一般指的是 whether、if、that。

【例】Whether an organism is a plant or an animal sometimes taxes the brain of a biologist.

一种生物究竟是植物还是动物，有时使生物学家颇伤脑筋。

2. 形式主语（It）＋谓语动词＋从句（由 that 或 whether 引导的从句是真正的主语）

这样的句子一般译为无人称句，如果先译从句，那么在翻译主句时，一般主句前加译"这"。

【例】It is a matter of common experience that bodies are lighter in water than they are in air.

物体在水中比在空气中轻，这是一种大家共有的经验。

在科技英语中还有一些与此相似的结构，如"It is (universally) known

that..." "It is believed that..." 等。而在翻译时，为了使译文成分完整，译者可以在句中加上泛指的主语（如人们、大家），译为："大家都知道……""人们都相信……"等。

（二）表语从句

表语从句用来说明主语的身份、性质、品性、特征和状态等。引导表语从句的词有连词、连接代词、连接副词和关系代词，例如 that、whether、why、how、when、where、what、as if、as though 等。通常按照句子的顺序，先翻译主句，再翻译从句。

【例】The result of invention of steam engine was that human power was replaced by mechanical power.

蒸汽机发明的结果是机械力代替了人力。

另外，在科技英语中还有一些结构如"That (This) is why...", "This (It) is because..." 等，译者可以按照字面直接译成"这就是为什么……""这就是为什么……的原因""这就是……的缘故"等。

（三）宾语从句

宾语从句一般可以分为 3 种：动词引导的宾语从句、介词引导的宾语从句和形容词引导的宾语从句，翻译时，一般顺序不变。

【例】1. Manure supplies what is deficient in the soil.

肥料供给土壤所缺乏的成分。

2. The ultimate disposal of these treated waste waters is, however, a function of whatever local, state and federal regulations may apply in any situation..

但是，这些处理过的废水最终如何处理，决定于在任何情况下都适用的地方、州或（美国）联邦的立法条例。

英语中引导宾语从句的形容词一般有 sorry、certain、sure、glad、happy、satisfied、surprised 等，一般不出现在科技英语当中。

二、形容词性从句

具有形容词功能，在复合句中做定语的从句被称为形容词性从句或定语从句。被修饰的名词、词组或代词被称为先行词。形容词性从句分为两种类型：由关系代词 who、whom、whose、that、which、as 引导的从句和由关系副词 when、where、why 引导的从句。

（一）逆序合译法

一般情况下，无论原文的内容长短，只要在汉译后放在被修饰语之前通顺,从句的翻译就可以置于修饰语之前。

【例】Stainless steel, which is very popular for its resistance to rusting, contains large percentage of chromium.

具有突出防锈性能而为大众所喜爱的不锈钢含铬的百分比很高。

(二) 顺序分译法

与上面情况相反,汉译时放在被修饰语之前不通顺,从句的翻译就应该后置,作为词组或分句。

【例】Each kind of atom seems to have a definite number of "hands" that it can use to hold on to others.

每一种原子似乎都有一定数目的"手",用来抓牢其他原子。(顺序分译法)

每一种原子似乎都有一定数目用于抓牢其他原子的手。(逆序合译法)

这句限制性定语从句虽然不长,但用顺序分译法译出的译文要比用逆序合译法更为通顺。定语从句较长,与主句关联又不紧密,汉译时就作为独立句放在主句之后。这种译法仍然是顺序分译法。

【例】Such a slow compression carries the gas through a series of states, each of which is very nearly an equilibrium state and it is called a quasi-static or a "nearly static" process.

这样的缓慢压缩能使这种气体经历一系列的状态,但各状态都很接近于平衡状态,所以称为准静态过程,或"近似稳定"过程。

(三) 拆译法

There + be 句型中的限制性定语从句汉译时往往可以把主句中的主语和定语从句融合一起,译成一个独立的句子。

【例】There are bacteria that help plants grow, others that get rid of dead animals and plants by making them decay, and some that live in soil and make it better for growing crops.

有些细菌能帮助植物生长,另一些细菌则通过腐蚀来消除死去的动物和植物,还有一些细菌则生活在土壤里,使土壤变得对种植庄稼更有好处。

三、副词性从句

副词性从句也称为状语从句,主要用来修饰主句或者主句的谓语,表示时间、原因、条件、让步等。

(一) 时间关系的译法

时间关系主要有 3 种:当时("在…之时")、先时("在…之前")和后时("在…之后")。另外还有限制时("每当…")和延续时("到…为止")等。如同数字、条件等，时间也是科技文献的重要数据。在翻译时要忠于原文，准确表达。准确不等于逐词死译，而是要紧扣原文内涵，以地道的英语加以翻译，必要时做一些变通处理。

1. 当时的译法

【例】1. Our results, however, indicated that no increase of 5-HT released into the ventricles could be found when morphine exerted its analgesic action.

我们的结果说明吗啡发挥镇痛作用时释放到脑室的 5-HT 并未增加。

2. This allows the two parts of the truck to be at an angle to each other when cornering.

这就使得这种卡车的两部分在转弯时彼此成一个角度。

3. Other things being equal, iron heats faster than aluminium.

其他情况一样时，铁比铝热得快。

2. 先时的译法

【例】1. Before the pollen grains were isolated from the anther, the anthers had been precultured on the H-medium for 1–4 days with the unprecultured anthers as the controls.

在分离花粉培养以前先将花药放在 H 培养基上进行 1~4 天的预培养，以未预培养的花药做对照。

2. The field must be well dug previous to the planting.

在播种之前，必须把土翻好。

3. 后时的译法

【例】1. The pollen grains had been induced to develop into cotyledon-shaped embryos, before they were transferred onto solid medium.

待花粉发育到子叶形胚状体后转移到固体培养基上。（"... before..." 结构常翻译为 "…之后才…"）

2. Chromium having been added, strength and hardness of the steel increased.

加入铬之后，钢的强度和硬度都增加了。

（二）条件关系

条件关系通常译成条件状语。此外，还可以用定语从句、含条件意义的时间状语从句、介词短语、分词短语等表达或做变通处理。

英语中的条件句分为真实的和虚拟的两种。前者表示的条件是真实的，后者表示的条件是假设的，或者是虚构的。

1. 真实的条件句

文中所叙述的条件是真实的，可以实现的。除 if 引导的从句表示条件以外，还有其他一些表示条件的表达方式。

【例】1. If an exhaust pipe is attached to a road vehicle it will always be fitted with a silencer.

如果排气管是安装于道路车辆的话，则总是配以消音器。

2. Any body above the earth will fall unless it is supported by an upward force equal to its weight.

离开地面的任何物体如果没有受到一个大小与其重量相等的力支持，就会掉下来。

2. 虚拟条件句

【例】If the small shoots were not transferred soon after differentiation, some abnormal phenomena might occur.

胚状体形成小苗后如不及时转移，往往会出现一些不正常现象。

3. 隐含译法

译成隐含条件关系的定语从句、以 when 引导的状语从句、分词短语以及其他结构。条件关系还可隐含在主谓结构中，这时主句一般表示未成的事实。

【例】1. The buyers are entitled to ask for replacement of the goods, which are found defective.

如果货物有毛病，买主有权要求更换。

2. When CH was substituted for YE, the callus turned yellow.

如果将 CH 换成 YE，则愈伤组织呈黄色。

(三) 原因关系的译法

原因关系在语言上的表达方式很多，而且不一定限于一个句子当中。但是这里所指的译法一般只限于一个句子当中所叙述的原因。常用的译法有原因状语从句、介词短语、分词短语、定语从句以及其他结构。翻译时应根据原因意义的强弱和语气轻重的不同加以选择。

1. 原因状语从句

【例】Because sulfur compounds are present in coal, crude oil, and gas, some sulfur dioxide is liberated when these burn.

因为硫化物存在于煤、原油以及煤气中，当这些东西燃烧时就会放出一些二氧化硫。

2. 介词短语

【例】Owing to the lack of endosperm in mature citrus seeds, we began to observe systematically the development of seeds two months after flowering by total dissection.

由于柑橘的成熟种子是没有胚乳的,故从开花后两个月开始,整体解剖观察,追踪种子的发育过程。

3. 分词短语或分词独立结构

【例】1. Being unable to reason correctly, scientists frequently fall into avoidable error.

科学家往往由于未能进行正确的推理而犯本来可以避免的错误。

2. Large sums of money being involved, it is necessary to direct these human and material resources into a specific channel with clear-defined objectives.

由于耗资巨大,须将人力物力纳入一个目标十分明确的既定渠道。

4. 隐含原因意义的定语从句、并列句或主谓结构等

【例】1. Hence, on this point, what we needed was a control experiment to rule out the nonspecific effect of DNA on embryogeneses, which had been known to exist for many years.

因而,在此地我们需要的是一个对照试验以排除DNA对胚胎发育的非特异性效应,因为多年来我们一直知道有这种效应的存在。

2. The increasing magnitude and complexity of the problems to be solved led to the setting up of large research teams.

因为要解决的问题日益增多而且复杂,所以建立了庞大的科研队伍。

(四) 让步关系的译法

让步关系就是从句先"退让"一步,认可事实,然后转入主句,摆脱这种事实的限制,实际意义是"进一步",语气有所加强。英语中让步关系的表达方式很多,翻译时根据情况加以选择,必要时做变通处理。

1. 让步状语从句

【例】Although the durations of perfusion varied, the results were almost the same.

灌流时间虽不同,但结果基本一致。

2. 让步关系的介词短语

【例】There are typical thyroid crisis manifestations and morality remains very high despite active management.

有典型甲亢危象表现,虽积极抢救,但死亡率极高。

3. 隐含让步关系的结构

【例】Calli continued to grow, but shoot differentiation never took place.

愈伤组织虽能继续生长，但始终没有分化出苗。

科技英语摘要写作（三）

指示-信息型摘要

指示-信息型摘要也可以称为指示-报道型摘要，为大部分杂志和论文集中所采用的摘要，既不是单纯指示型的也不是单纯信息型的，而是将二者有机地结合起来，将原文献中信息价值高的部分写成信息型的摘要，其余部分则写成指示型摘要，兼具二者特点。

这种类型的摘要在信息部分浓缩了文章中的主要精华，信息量大，而且不加入评论和解释，客观准确地表达出原文的重要内容。其余部分则以指示型摘要形式表达，具有独立性和完整性，可以有效地起到传递信息、检索、参考和交流的功能，为读者节省阅读时间，避免因语言障碍所导致的信息检索困难。

值得注意的是，和其他两个类型的摘要一样，指示-信息型摘要行文要结构严谨，表达简明，语义确切，逻辑性强，上下连贯，互相呼应。同时在摘要中，要谨慎使用长句，句型应力求简单。表意明白，无空泛、含混之词。

在论文尤其是科技论文写作中，往往会有很多术语或符号出现，因此在摘要中要特别注意术语的规范化，避免使用非公知公用的术语和符号。新术语或还没有准确表达的术语，可用原文或译文后加括号注明。在这种摘要中一般不用插图、表格、数学公式、化学结构式等。

【范文】**Epidemiological Survey of Graves' Disease on a Hundred Thousand People in Daqing Area**

Abstract: In this paper, we studied the incidence of Graves' disease and associated factors. We did so by conducting epidemiological investigations about a total of 100,123 people aged 15-59 in Daqing region for an average of two years. We started by surveying 301 patients with Graves' disease who were newly diagnosed, 71 of whom (0.16%) were males and 227 (0.41%) females. The total incidence was 0.3%. Differences of incidence between ages and occupations were significant. Based on the investigations of 18 associated factors for Graves' disease, the Logistic multivariate regression analysis was used to analyze the results. It showed the following pathological factors: virus history, psychological stimulation, food, drugs, family history, education, distributions, etc. Finally, we conclude that this study may provide theoretical evidence for prevention and treatment of Graves' disease.

摘要译写示例

示例一

【摘要中文原文】

摘要：为有效提高猪血的利用开发率，本研究采用液体发酵方法，以猪血水解度为测定指标，从屠宰场污水池中分离得到1株降解猪血蛋白能力较强的菌株 X-17，并采用单因素试验和正交试验对 X-17 进行培养基和发酵条件进行优化。获得其最佳培养基配方为：玉米粉 1%，磷酸氢二钠 0.1%，硫酸镁 0.05%；最优发酵条件为：接种量 4%，培养温度 37℃，培养基起始 pH 为 7.0，发酵时间 60h。X-17 通过最优条件发酵猪血后，水解度达到了 18.52%，可溶性蛋白提高了 181%，氨基态氮增加 106%。

【摘要原版译文】

Abstract: In order to effectively improve the swine blood, the study was used by liquid fermentation method, and uses the DH of fermented liquid as index, a bacillus X-17 could degradation protein of swine blood in fermentation liquid that was screened from the cesspool. Culture medium and fermentation conditions of bacillus X-17 were optimized by the single factor and orthogonal experiment. Ultimately to determine the optimal medium composition: 1% cornmesl, 0.1% Na_2HPO_4, 0.5% $MgSO_4$, temperature 37, initial pH 7.0, fermentation time 60 h. Bacillus X-17 was fermented by optimal conditions, degree of hydrolysis in fermentation liquid was up to 18.41%, and dissolving protein increased 181%, the amino nitrogen increased 106%.

【摘要修改译文】

Abstract: In order to effectively improve utilization of swine blood, by liquid fermentation method, using DH of fermented liquid as index, a strain X-17 with strong ability to degrade protein of swine blood was screened from the cesspool of a slaughterhouse. Culture medium and fermentation conditions of strain X-17 were optimized by single factor and orthogonal experiment. The optimal medium compositions obtained are: 1% cornmeal, 0.1% Na_2HPO_4, 0.05% $MgSO_4$; and the optimal fermentation conditions are: amount of inoculation 4%, culture temperature 37 ℃, initial pH of culture medium 7.0, fermentation time 60 h. Strain X-17 was fermented with the optimal conditions, degree of hydrolysis in fermentation liquid reached 18.52%, and soluble protein increased by 181% and amino nitrogen increased by 106%.

【主要修改意见】

1. 译文有不忠实于原文的现象。例如第一句中汉语"为提高猪血的利用开发率"被翻译为"In order to effectively improve the swine blood",漏译"利用开发率",这样的译文令人费解,因此改为"In order to effectively improve utilization of swine blood"。

2. 被动语态是科技英语写作和翻译中的一个显著特点,无明显人称或相应成分做主语的句子应该用被动语态。原文中"获得其最佳培养基配方为…"没有明显主语但译文中应该体现出来(The optimal medium compositions obtained are…),而不是按字面表达直接翻译为"Ultimately to determine the optimal medium composition"。

3. 用词不准确。原文中"提高了"需译为"increased by","提高到"译为"increased to",而不是仅仅用动词"increase"表达。

示例二

【摘要中文原文】

摘要:淀粉是禾谷类作物籽粒中的主要储藏化合物,广泛应用于化工、医药、纺织、造纸、建筑等领域。随着淀粉需求量的急剧增加,如何提高作物淀粉含量及改良淀粉品质是各个领域研究的热点。基因工程技术改良作物具有短时、高效的特点,是目前培育高淀粉品种的重要手段。淀粉合成是个复杂的代谢过程,受多种酶的调节,启动子作为基因表达的重要调控元件,在淀粉合成中起着至关重要的作用。随着基因工程的发展,经常需要高活性、特异性驱动异源蛋白表达的载体,选择合适的启动子是驱使外源基因高效、特异表达的关键,也是培育安全转基因作物的首要问题。本文综述了淀粉合成相关酶启动子的研究进展,旨在为从基因调控水平上提高淀粉含量及改良淀粉品质等方面的研究提供理论参考。

【摘要原版译文】

　　Abstract: Starch is the main storage compound of the cereal in the grain crops, which is widely used in chemical, pharmaceutical, textile, papermaking, and construction fields. With the dramatic increase in demand for starch, how to increase the content of starch and improve the quality of starch in crops has become the hot topics in the each research fields. The crop improvement is an important means to cultivate high-starch varieties currently by genetic engineering techniques that it have short and efficient features. Starch synthesis is a complex metabolic processes regulating by a variety of enzymes, and that the promoter is an important regulatory elements of gene expression, which plays a vital role in starch synthesis. With the developing of genetic engineering, there often need more heterologous expression vectors that can express a high level or specificity. therefore, how to choose a appropriate promoter driven the exogenous gene expression that is the key and it is also the chief problem for cultivating safety transgenic plants. In this paper, the research progress of the starch synthesis-related enzyme promoter were reviewed to provide the theory reference that raising the content of starch and modifying the quality of starch in the level of gene regulation.

【摘要修改译文】

　　Abstract: Starch is the main storage compound of cereal in grain crops, which is widely used in chemical, pharmaceutical, textile, papermaking, and construction fields. With the dramatic increase in demand for starch, the ways to increase content of starch and improve quality of starch in crops have become hot issues in each research field. Genetic engineering technology is an important means currently used to breed high-starch varieties which possess short-time and high-efficiency features. Starch synthesis is a complex metabolic process regulated by a variety of enzymes, and promoter is an important regulatory element of gene expression, which plays a vital role in starch synthesis. With the development of genetic engineering, more heterologous expression vectors that can express a high level or specificity are often needed. Therefore, choosing an appropriate promoter is the key to drive exogenous gene to express with high efficiency and specificity and it is also the chief issue for cultivating safety transgenic plants. In this paper, research progress of starch synthesis-related enzyme promoter was reviewed to provide theoretical reference for researches on increasing content of starch and improving quality of starch from the level of gene regulation.

【主要修改意见】

1. 原版译文中有明显细节错误。如"with the developing of genetic engineering",另外,句首单词首字母不大写,如原译文中句子开头的"therefore"等,这些在翻译中由于疏忽造成的问题应该在校对时及时发现并改正。

2. 原版译文中有明显的中式英语表达。如"基因工程技术改良作物具有短时、高效的特点"译成"genetic engineering techniques that it have short and efficient features",而且从句中还多出一个主语"it"。

3. 在翻译过程中,译者首先应该把原文的信息清楚、准确地传递给读者,而不只是注重语言的形式是复杂还是简单,因此并不是句式越复杂,从句越多越好。原版译文中,译者使用的多处从句导致原文信息没有被清楚地表达出来,如"如何选择合适的启动子是驱使外源基因高效、特异表达的关键"译成了从句"how to choose a appropriate promoter driven the exogenous gene expression that is the key",结构错误且语义不清楚,再如最后一句"旨在为从基因调控水平上提高淀粉含量及改良淀粉品质等方面的研究提供理论参考"译成了从句"provide the theory reference that raising the content of starch and modifying the quality of starch in the level of gene regulation",结构也是错误的。

Unit 4   Environment

Text A

Water, Agriculture and the Environment

Crops and fodder need water to grow. As a rough indication, a kilogram of potatoes takes up to 500 liters of water to produce, a kilogram of grain-fed beef up to 100,000. Throughout most of the world, and in many rural areas of the OECD, agriculture is the leading consumptive user of water. How are OECD governments responding to the increased demand for water in agricultural production and the growing public awareness of the impact of agriculture on the environment?

Agriculture uses a lot of water. Withdrawals for agricultural purposes currently account for 50% or more of total abstraction in at least nine OECD countries (Greece, Italy, Japan, Korea, Mexico, New Zealand, Portugal, Spain and Turkey), and for over a third in five others (Australia, Denmark, Hungary, the Netherlands and the United States). But because of losses through evaporation and plant transpiration, the share of agriculture in total water consumption is usually much higher. In North America and OECD Europe, for instance, it accounts for between 37 and 45% of total withdrawals, but around two-thirds of consumption. In drier parts of the OECD area, such as California, and the Murray-Darling basin in Australia, agriculture accounts for well over 90% of consumptive use.

Irrigation has allowed the expansion of agriculture into semi-arid and even arid environments. With the exception of crops with low value-added (like alfalfa), irrigation can bring substantial economic gains. As a rule, farmers have free access to (or are charged only a nominal fee for) water that they pump themselves. And several countries (among others, Mexico, Turkey and the United States, at least in some federal irrigation districts) continue to offer preferential tariffs for electricity used to pump water for irrigation. Nowadays, governments in most OECD counties are scaling back large programmers to develop irrigation (Turkey is the major exception), as they run up against physical, if not financial, limits. The annual rate of expansion of irrigated area in the OECD has dropped to below 1% (compared with close to 3% in the late 1970s). But the exploitation of groundwater aquifers for farming continues to grow apace—not only in arid and semi-arid regions but in many humid areas as well. In the United Kingdom, for example, the volume of water abstracted for irrigation has more than tripled in the last 25 years, although from a small base.

Water-use in agriculture has marked effects on the environment, not all of them adverse. Reservoirs created for irrigation can provide fresh water for birds and other fauna; terraces for growing rice can help slow down runoff and reduce erosion; water-management for agricultural purposes can replenish the water-table and stabilize river levels. Irrigation, moreover, allows the recycling of urban waste-waters—although such irrigated, its share of the gross value of agricultural production is somewhere in the neighborhood of 40%. To some extent irrigation has also helped with risk-management, thus reducing pressures on government to provide disaster payments to compensate for crop losses as a result of periodic droughts. The major share of the costs of investments in irrigation falls on the taxpayer, other water-users and electricity consumers (through cross-subsidies). And it is, in the main, national treasuries that have financed dams, reservoirs and delivery networks as well as a large part of the cost of installing local and farm infrastructure, including drainage pipes. Governments generally attempt to recover some of these costs through user-charges, but revenues are rarely enough to cover even operation and maintenance costs. The economic distortions caused by the often enormous underpricing of surface water used in agriculture have been compounded in many instances by agricultural policies, particularly those linked to the production of particular commodities.

The diversion of water for agriculture has often caused environmental problems and degraded natural resources. The steady flows that dams cause alter the seasonal cycles of aquatic and riparian plants and animals. Reduced stream flows and excessive use of groundwater aquifers lead to higher concentration of pollutants. Excessive extraction can lower water-tables, leading, in some cases, to ground subsidence and, in some coastal areas, to salt-water intrusion. And because irrigation water almost always contains much higher concentrations of dissolved salts than rainwater, its discharge often raises the proportion of salts in the bodies of water into which it flows. The depletion of aquifers for irrigation raises questions about the sustainability of some important farming systems. In the Texas High Plains of the United States, a major cotton, grain and beef production area, agriculture has been responsible for depleting one-quarter of the Texas portion of the massive Ogallala aquifer—an underground water body containing the run-off of several ice-ages—over the five decades to 1990.

Another threat to the sustainable management of agricultural land is the lack of adequate drainage, which farmers and governments also fail to provide because of its expense. The result too often is water-logging (the over-saturation of soils with water) and the build-up of salt in soils. In places such as the Iberian Peninsula, Australia and western North America, fertile lands have had to be abandoned due to salt encrustation, nullifying some of the gains that irrigation was intended to yield. Problems arising from irrigation are not only ones associated with water-management practices in agriculture. In Queensland, Australia, for example, the concentration of cattle around watering stations fed by artesian wells has degraded grazing land. In the Netherlands and in parts of other OECD countries with low-lying farmland, the drainage of water to facilitate cultivation has had a profound effect on the flora and fauna. Generally, species adapted to wet conditions die out, to be displaced by those that prefer drier soils. Adding to the general concern over the sustainable use of water are the uncertainties about the possible effects of global climate change. If temperatures were to rise substantially as a result of an enhanced greenhouse effect, both rain-fall patterns and evaporate-transpiration rates are likely to be affected: some areas of the OECD, such as northern Europe, could experience increases in rainfall, and others, such as Western Australia and Japan, could face declines. Clearly, any such changes would have major implications for agriculture, water and the environment.

(1,021 words)

New Words

1. fodder ['fɔdə] *n.* 饲料
2. consumptive [kən'sʌmptiv] *adj.* 消耗性的
3. abstraction [æb'strækʃən] *n.* 抽取
4. evaporation [i,væpə'reiʃən] *n.* 蒸发
5. transpiration [,trænspi'reiʃən] *n.* 蒸腾（作用）
6. basin ['beisn] *n.* 盆地；流域
7. arid ['ærid] *adj.* 干燥的；干旱的
8. alfalfa [æl'fælfə] *n.* （植物）苜蓿
9. substantial [səb'stænʃəl] *adj.* 大量的
10. nominal ['nɔminəl] *adj.* 微不足道的；名义上的
11. aquifer ['ækwifə] *n.* 地下蓄水层
12. apace [ə'peis] *adv.* 迅速地
13. abstract [æb'strækt] *vt.* 抽取
14. adverse ['ædvə:s] *adj.* 不利的；有害的
15. reservoir ['rezəvwɑ:] *n.* 水库；蓄水池
16. fauna ['fɔ:nə] *n.* 动物区系
17. terrace ['terəs] *n.* 梯田
18. runoff ['rʌnɔf] *n.* 地表径流
19. erosion [i'rəuʒən] *n.* 侵蚀；腐蚀
20. replenish [ri'pleniʃ] *vt.* 补充，再装满
21. stabilize ['steibilaiz] *vt.* 使稳定
22. periodic [,piəri'ɔdik] *adj.* 周期的；定期的
23. drainage ['dreinidʒ] *n.* 排水；排水系统
24. revenue ['revənju:] *n.* 收入；收益；税收
25. maintenance ['meintənəns] *n.* 维护；维修
26. distortion [dis'tɔ:ʃən] *n.* 扭曲
27. diversion [dai'və:ʃən] *n.* 转移
28. aquatic [ə'kwætik] *adj.* 水生的；水栖的
29. riparian [rai'pɛəriən] *adj.* 河边的；水滨的
30. concentration [,kɔnsən'treiʃən] *n.* 浓度；浓缩

31. intrusion [in'tru:ʒən] *n.* 侵入；打扰
32. discharge [dis'tʃɑ:dʒ] *n.* 排出；流量
33. depletion [di'pli:ʃən] *n.* 损耗；用尽
34. water-logging ['wɔ:tə'lɔgiŋ] *n.* 渍涝
35. saturation [,sætʃə'reiʃən] *n.* 饱和
36. encrustation [in,krʌs'teiʃən] *n.* 结壳
37. nullify ['nʌlifai] *vt.* 取消
38. artesian [ɑ:'ti:zjən] *adj.* 自动流出的
39. grazing ['greiziŋ] *n.* 放牧；牧场
40. cultivation [,kʌlti'veiʃən] *n.* 耕种，耕作；栽培
41. flora ['flɔrə] *n.* 植物区系

Useful Expressions

1. as a rule   一般说来
2. have access to   使用
3. preferential tariff   特惠关税
4. scale back   按比例缩减
5. run up against   遇到
6. in the neighborhood of   大约
7. farm infrastructure   农田基本建设
8. ground subsidence   地面沉陷
9. dissolved salts   溶解盐类

Notes

1. 本文选自《资源环境科学专业英语》，中国农业出版社，2008，作者：许修宏。

2. 经济合作和发展组织（Organization for Economic Cooperation and Development，OECD），是由34个市场经济国家组成的政府间国际经济组织，旨在共同应对全球化带来的经济、社会和政府治理等方面的挑战，并把握全球化带来的机遇。成立于1961年，总部设在巴黎。

3. 蒸发（evaporation），是物质从液态转化为气态的相变过程。

4. 蒸腾（transpiration），是水分从活的植物体表面（主要是叶片）以水蒸气状态散失到大气中的过程。与物理学的蒸发过程不同，蒸腾作用不仅受外界环境条件的影响，而且还受植物本身的调节和控制，因此它是一种复杂的生理过程。

5. 地下水位（water-table），即地下含水层中水面的高度。根据钻探观测时间可分为初见水位、稳定水位、丰水期水位、枯水期水位、冻前水位等。

6. 渍涝（water-logging），是一种气象灾害，一般情况而言，单次暴雨过程不易形成渍涝灾害，渍涝灾害往往是由连续性的较大降水造成的。

Exercises

Part Ⅰ　Vocabulary and Structure

Section A　Match each word with its Chinese equivalent.

1. sewage　　　　　　　　A. 堆肥
2. composting　　　　　　B. 荒漠化
3. desalinization　　　　　C. 污水
4. dilution　　　　　　　　D. 农业生物多样性
5. leachate　　　　　　　　E. 富营养化
6. eutrophication　　　　　F. 沥出液
7. duststorm　　　　　　　G. 稀释
8. afforestation　　　　　　H. 脱盐
9. agrobiodiversity　　　　 I. 尘暴
10. desertification　　　　　J. 绿化造林

Section B　Fill in the blanks with the words or expressions given below. Change the form where necessary.

arid	replenish	evaporate	erosion	concentrate
drain	aquatic	encrustation	aquifer	deplete

1. Increased consumption of water has led to rapid _____ of groundwater reserves.

Unit 4　Environment

2. There are extreme _____ regions where only the fittest survive.
3. Moisture is drawn to the surface of the fabric so that it _____ .
4. Untrammeled industrialization, particularly in the poor countries, is contaminating the river and _____ .
5. The pond is small but can support many _____ plants and fish.
6. Although all plants normally contain fluorine（氟）, the _____ varies greatly.
7. Your body is in survival mode and needs food to _____ its energy.
8. The disrupted _____ will cause whole hillsides of waterlogged or dried out peat（泥炭）to slide.
9. The products may serve to protect the water tanks, engines and metal pieces, and to prevent _____ and raise heating rate, more fuel-efficient energy.
10. The _____ rain is defined as rain, which can cause soil erosion.

Part Ⅱ　Translation

Section A　Translate the following sentences into Chinese.

1. Reduced stream flows and excessive use of groundwater aquifers lead to higher concentrations of pollutants.
2. Water-use in agriculture has marked effects on the environment, not all of them adverse.
3. It is, in the main, national treasuries that have financed dams, reservoirs and delivery networks as well as a large part of the cost of installing local and farm infrastructure, including drainage pipes.
4. The diversion of water for agriculture has often caused environmental problems and degraded natural resources.
5. Another threat to the sustainable management of agricultural land is the lack of adequate drainage.

Section B　Translate the following sentences into English.

1. 池塘和湖泊在陆地上都是相对暂时的景象，因为它们最终将会被淤泥（silt）所填平。
2. 细菌是土壤中数量最大的微生物。的确，它们是地表上最常见的生命有机体。

3. 然而，污染（contamination）可能沿着从田地到餐桌的这条食物链，在其任何一点上发生。
4. 随着世界气候的变迁，恐龙绝迹了，但许多较小的动物却继续活了下来。这就是适者生存。
5. 那就意味着被药物污染的废水会进入地下水和地表水，而这正是大部分人饮用水的主要来源。

Text B

What on Earth Are We Doing?

At the outside limit, the earth will probably last another 4 billion to 5 billion years. By that time, scientists predict, the sun will have burned up so much of its own hydrogen fuel that it will expand and *incinerate* （烧成灰） the surrounding planets, including the earth. A nuclear *cataclysm* （大灾难）, on the other hand, could destroy the earth tomorrow. Somewhere within those extremes lies the life expectancy of this wondrous, swirling globe. How long it endures and the quality of life it can support do not depend alone on the *immutable* （不可变的） laws of physics. For man has reached a point in his evolution where he has the power to affect, for better or worse, the present and future state of the planet.

Through most of his 2 million years or so of existence, man has thrived in earth's environment—perhaps too well. By 1800 there were 1 billion human beings bestriding the planet. That number had doubled by 1930 and doubled again by 1975. If current birthrates hold, the world's present population of 5.1 billion will double again in 40 more years. The frightening irony is that this *exponential* （指数的） growth in the human population—the very sign of *Homo sapiens*' success as an organism—could doom the earth as a human habitat.

The reason is not so much the sheer numbers, though 40,000 babies die of starvation each day in Third World countries, but the reckless way in which humanity has treated its planetary host. Like the evil genies that flew from Pandora's box, technological advances have provided the means of upsetting nature's equilibrium, that intricate set of biological, physical and chemical interactions that make up the web of life. Starting at the dawn of the Industrial Revolution, smokestacks have disgorged *noxious*（有毒的）gases into the atmosphere, factories have dumped toxic wastes into rivers and streams, automobiles have guzzled irreplaceable fossil fuels and fouled the air with their *detritus*（碎屑）. In the name of progress, forests have been *denuded*（使光秃）, lakes poisoned with pesticides, underground aquifers pumped dry. For decades, scientists have warned of the possible consequences of all this *profligacy*（肆意挥霍）. No one paid much attention.

This year the earth spoke, like God warning Noah of the *deluge*（洪水）. Its message was loud and clear, and suddenly people began to listen, to ponder what portents the message held. In the U.S., a three-month drought baked the soil from California to Georgia, reducing the country's grain harvest by 31% and killing thousands of head of livestock. A stubborn seven-week heat wave drove temperatures above 100°F across much of the country, raising fears that the dreaded "green-house effect"—global warming as a result of the buildup of carton dioxide and other gases in the atmosphere—might already be under way. *Parched*（烤干）by the lack of rain, the Western forests of the U.S., including Yellowstone National Park, went up in flames, also igniting a bitter conservationist controversy. And on many of the country's beaches, garbage, raw sewage and medical wastes washed up to spoil the fun of bathers and confront them personally with the growing *despoliation*（掠夺）of the oceans.

Similar pollution closed beaches on the Mediterranean, the North Sea and the English Channel. Killer hurricanes ripped through the Caribbean and floods devastated Bangladesh reminders of nature's raw power. In Soviet Armenia a monstrous earthquake killed some 55,000 people. That too was a natural disaster, but its high casualty count, owing largely to the construction of cheap high-rise apartment blocks over a well-known fault area illustrated the carelessness that has become humanity's habit in dealing with nature.

There were other *forebodings*（先兆）of environmental disaster. In the U.S. it was revealed that federal weapons-making plants had recklessly and secretly littered large areas with radioactive waste. The further depletion of the atmosphere's ozone layer, which helps block cancer-causing ultraviolet rays, testified to the continued overuse of atmosphere-destroying chlorofluorocarbons *emanating* (发出) from such sources as spray cans and air-conditioners. Perhaps most **ominous** of all the destruction of the tropical forests home to at least half the earth's plant and animal species, continued at a rate equal to one football field a second.

Most of these evils had been going on for a long time, and some of the worst disasters apparently had nothing to do with human behavior. Yet this year's bout of freakish weather and environmental horror stories seemed to act as a powerful catalyst for world-wide public opinion. Everyone suddenly sensed that this gyrating globe, this precious *repository* (仓库) of all the life that we know of, was in danger. No single individual, no event, no movement captured imaginations or dominated headlines more than the clump of rock and soil and water and air that is our common home. Thus in a rare but not unprecedented departure from its tradition of naming a Man of the Year, TIME has designated Endangered Earth as Planet of the Year for 1988.

What would happen if nothing were done about the earth's imperiled state? According to computer projections, the accumulation of CO_2 in the atmosphere could drive up the planet's average temperature 3°F to 9°F by the middle of the next century. That could cause the oceans to rise by several feet, flooding coastal areas and ruining huge tracts of farmland through *salinization* (盐碱化). Changing weather patterns could make huge areas infertile or uninhabitable, touching off refugee movements unprecedented in history.

Toxic waste and radioactive contamination could lead to shortages of safe drinking water, the *sine qua non* (必要条件) of human existence. And in a world that could house between 8 billion and 14 billion people by the mid-21st century, there is a strong likelihood of mass starvation. It is even possible to envision the world so wryly and chillingly *prophesied* (预言) by the typewriting cockroach in Donald Marquis' *Archy and Mehitabel*: "Man is making deserts of the earth/it won't be long now/before man will have it used up/so that nothing but ants/and centipedes and scorpions/can find a living on it."

Humanity's current predatory relationship with nature reflecting a man-centered world view has evolved over the ages. Almost every society has had its myths about the earth and its origins. The ancient Chinese depicted Chaos as an enormous egg whose parts separated into earth and sky, yin and yang. The Greeks believed Gaia, the earth, was created immediately after Chaos and gave birth to the gods. In many *pagan*（异教徒的） societies, the earth was seen as a mother, a fertile giver of life. Nature—the soil, forest, sea—was endowed with divinity, and mortals were subordinate to it.

The Judeo-Christian tradition introduced a radically different concept. The earth was the creation of a monotheistic God, who, after shaping it, ordered its inhabitants, in the words of Genesis: "Be fruitful and multiply, and replenish the earth and subdue it; and have dominion over the fish of the sea and over the fowl of the air and over every living thing that moveth upon the earth." The idea of dominion could be interpreted as an invitation to use nature as a convenience. Thus the spread of Christianity, which is generally considered to have paved the way for the development of technology, may at the same time have carried the seeds of the *wanton*（挥霍的） exploitation of nature that often accompanied technical progress.

Those tendencies were compounded by the Enlightenment notion of a mechanistic universe that man could shape to his own ends through science. The exuberant optimism of that world view was behind some of the greatest achievements of modern times: the invention of laborsaving machines, the discovery of anesthetics and vaccines, the development of efficient transportation and communication systems. But, increasingly, technology has come up against the law of unexpected consequences. Advances in health care have lengthened life-spans, lowered infant-mortality rates and, thus, *aggravated*（加重） the population problem. The use of pesticides had increased crop yields but polluted water supplies. The invention of automobiles and jet planes has revolutionized travel but *sullied*（玷污） the atmosphere.

Let there be no illusions. Taking effective action to halt the massive injury to the earth's environment will require a mobilization of political will, international cooperation and sacrifice unknown except in wartime. Yet humanity is in a war right now, and it is not too Draconian to call it a war for survival. It is a war in which all nations must be allies. Both the causes and effects of the problems that threaten the earth are global, and they must be attacked globally. "All nations are tied together as to their common fate," observes Peter Raven, director of the Missouri Botanical Garden. "We are all facing a common problem, which is, how are we going to keep this single resource we have, namely the world, *viable*（能存活的）?"

As man heads into the last decade of the 20th century, he finds himself at a crucial turning point: the actions of those now living will determine the future, and possibly the very survival, of the species. "We do not have generations, we only have years, in which to attempt to turn things around," warns Lester Brown, president of the Washington-based Worldwatch Institute. Every individual on the planet must be made aware of its vulnerability and of the urgent need to preserve it. No attempt to protect the environment will be successful in the long run unless ordinary people—the California housewife, the Mexican peasant, the Soviet factory worker, the Chinese farmer—are willing to adjust their lifestyles. Our wasteful, careless ways must become a thing of the past. We must recycle more, procreate less, turn off lights, use *mass transit*（公共交通）, do a thousand things differently in our everyday lives. We owe this not only to ourselves and our children but also to the unborn generations who will one day inherit the earth.

(1,703 words)

Comprehension of the Text

Choose the best answer to each of the following questions.

1. According to the text, there will be _____ people on the earth in 40 more years if current birthrates hold.
   A. 5 billion   B. 5.1 billion
   C. 10 billion   D. 10.2 billion
2. Which of the following is not the reason for destroying our habitat according to the author?
   A. Technological advances.   B. Pandora's box.
   C. Smokestacks.   D. Fuel-consuming automobiles.
3. The following places except _____ are mentioned in the text for having suffered natural disaster in the recent time.
   A. the South China Sea   B. the North Sea
   C. the English Channel   D. the Mediterranean
4. The italicized and boldfaced word "ominous" in paragraph 6 could probably

Unit 4　Environment

   be replaced by _____ .
   A. fortunate   B. frightening
   C. underlying   D. unfavorable
5. Why has TIME designated Endangered Earth as Planet of the Year rather than named a Man of the Year for 1988?
   A. Because no celebrities are entitled to the honor.
   B. Because TIME expected to attract more readers.
   C. Because TIME hoped to arouse people's environment-protection awareness.
   D. Because so many worst disasters happened on earth that year.
6. The following except _____ would happen if nothing were done about the earth's endangered state.
   A. shortages of safe drinking water
   B. mass starvation
   C. rise of temperature
   D. discontinuity of natural resources
7. The author called for international cooperation against the massive damage to the earth's environment mainly because _____ .
   A. the causes and effects of the problem that threaten the earth are global
   B. it is very urgent to save the people in the undeveloped area
   C. it is closely related to every nation's political interest
   D. the outcome of the pollution has been realized by every country
8. According to the author, the attempt to protect the environment will be successful if _____ .
   A. each government takes actions
   B. ordinary people are willing to adjust their life styles
   C. more advanced technologies can be invented
   D. the Washington-based Worldwatch Institute takes its initiatives
9. The author pointed out that it was the _____ that guided people to overdevelop the earth's resources.
   A. ancient Chinese myths   B. Greek mythologies
   C. radical concept of Christianity   D. Islamism
10. The author's attitude towards environment-protection is _____ .
    A. positive   B. negative
    C. neutral   D. critical

科技文体翻译技巧（四）

特殊词的译法

科技工作者不断创造出新的词汇和术语来命名或者表达自己科研领域的发现或者发明，因而科技文体的一大特点体现在词汇层面就是科技词汇的大量使用。这些特殊的科技词汇是科技文体翻译活动的一大难点。如果没有相关的专业知识，不了解这些特殊词的构成规律，那么就会给翻译带来诸多不便，甚至导致误译。虽然科技词汇数量大、来源广泛，但它们并非无章可循。

一、特殊词的来源和构成

（一）由通用英语词汇转化而来的专业词汇

科技英语文体中某些专业词汇是在赋予通用英语词汇以特定的新意后产生的。这些词汇在通用英语中早已存在，科技工作者常根据自己的需要将这些词汇进行修改或加以限制，这就使得通用英语词汇在科技英语中成为具有特定含义和习惯用法的专业术语。例如 probability 在日常生活中表示"可能性"，在物理学中表示"几率"，在数学上表示"概率"。再如 carrier 的日常普通含义是"搬运工"，但在不同的专业领域中，它可以是：（计算机）媒体、（集成电路）载体、（半导体）载流子、（运输）搬运车、（航空）运输机、（航天）运载火箭、（军事）航空母舰、（医学）带菌体等。翻译这类词汇切不可望文生义，而应根据其所在学科领域和所处的上下文认真辨明其真实含义。

（二）合成构词法创造的新词汇

将两个或者两个以上词合并一起构成新词的方法称为合成法（compounding），主要有4种合成词的构词方法：合写式、连字符式、开放式以及混合式。合写式的如 greenhouse（温室）、Internet（因特网）；连字符式的如 salt-former（卤素）、full-enclosed（全封闭的）、work-harden（加工硬化）；开放式的如 mouse pointer（鼠标指针）；混合式就是将两个单词按一定的规律进行剪裁，两个词各取一部分加以叠合混成一个新的单词或取一个词的一部分加另一个词的原型加以拼缀而成一个新的单词。混合式在科技英语中频繁使用，例如 netizen（网民）= net ＋ citizen，smog（烟雾）= smoke ＋ fog，medicare（医疗保健）= medical ＋ care，blog（博客）= web ＋ log，copytron（电子复写技术）= copy ＋ electron 等。合成法构词在英式英语和美式英语中也有一定差异。在美式英语中,合成词一旦获得某种永久地位就通常采用合写式,否则就采用开放式;而在英式英语中,合成词书写时通常采用连字符式。

(三) 派生法扩展而来的特殊词汇

派生法也称为词缀法(affixation),即在原有的单词或词干的前面或后面通过加词缀来构成新词,这种词就称为派生词(derivative)。词缀具有极强的灵活性,同时又具有极强、极广泛的搭配表意能力。因为一方面词缀的基本词义都比较稳定、明确;另一方面它们的黏附力都很强,黏附在词根之前或之后,概念就可以立即形成。因此派生法是英语构词法的一种主要手段,也是构成科技词汇的一种重要方法,例如 bio-(生命、生物)、thermo-(热)、electro-(电)、aero-(空气)、carbo-(碳)、hydro-(水)、-ite(矿物)、-mania(热、狂)等。如果我们熟悉这些前缀和后缀,这样组成的许多科技词汇就可以得到恰当的翻译。以下是一些典型例子。

miniultrasonicprober = mini + ultra + sonic + prober 微型超声波金属探伤仪

macrospacetransship = macro + space + trans + ship 巨型空间转运飞船

deoxyribonucleic = de + oxy + ribo + nucleic 脱氧核糖核酸的

barothermograph = baro + thermo + graph 气压温度记录器

photomorphogenesis = photo + morpho + genesis 光形态发生

(四) 由缩略法构成的专业词汇

首字母缩略法即取某一词组中主要单词的第一个字母组成新词的方法。随着科技发展,越来越多的缩略词在不同领域中广泛使用。但是由于缩略词简单,容易产生歧义,所以译者要根据上下文分辨其具体含义,例如 GPS (global positioning system) 全球定位系统、MRI (magnetic resonance imaging) 磁共振成像、FM (frequency modulation) 调频、DNA (deoxyribonucleic acid) 脱氧核糖核酸、RAM (random access memory) 随机存取存储器、ADP (automatic data processing) 自动数据处理、AIDS (acquired immune deficiency syndrome) 艾滋病、SALT (strategic arms limitation talks) 限制战略武器会谈等。

(五) 借用外来词和其他学科的词汇所扩展的科技英语专业词汇

部分科技英语词汇来源于外来语。一些新出现的科技术语经常借用其他语言的词汇,其中拉丁语和希腊语对英语科技词汇的影响源远流长,例如 appendicitis(阑尾炎)、thermonuclear(热核)、microscope(显微镜)、acrophobia(恐高症)等。除此以外,英语科技词汇还不断借用其他语言来丰富和发展自身。例如借用法语的有 chiffon(雪纺绸),借用波斯语的有 bazaar

(市场)、借用德语的有 antibody（抗体），借用意大利语的有 fresco（壁画），借用日语的有 ikebana（插花艺术），借用马来语的有 paddy（稻）等。

二、特殊词的翻译技巧

（一）意译

意译就是对原词所表达的具体事物和概念进行仔细推敲，以准确译出该词的科学概念。许多通用英语词汇被借用到科技英语文体当中来，在翻译这样的词汇时，我们必须注意在上下文里词汇所表达的意义，不能望词生义，想当然地按照该词汇在通用英语表达中的含义来译。相反，我们应该根据其所存在的科技英语语境来构建它的意义。

【例】The emission of CO_2 must be reduced for the purpose of stabilizing the *concentration* of CO_2 at a safe level.

为了将二氧化碳的浓度稳定在一个安全的水平上，必须减少二氧化碳的排放。

concentration 的一般含义是"集中，专心"，在科技英语的意思是"浓度"。同一个英语单词在不同专业里会有不同译法，甚至不同专业的各种术语可能在同一文章中交叉出现。milking 在畜牧业中有"产奶"（milking rate 产奶率）的意思，在水稻生产中是"灌浆"（milking period 灌浆期）的意思。service 的一般含义为"服务"，在畜牧繁殖里却指"配种"，如 artificial service（人工繁殖）、service crate（配种架）。

上面的例子说明，我们在翻译科技文体的文章时，对于似是而非的单词应勤查专业词典，根据语境去审度词义。

（二）直译

文化差异性很少体现在科技文体中，同时科技词汇的意义也比较专一，所以对很多科技文体中的字、词采取直译的方法。如：

aerobic composting 好氧堆肥

environmental impact 环境影响

air quality standard（AQS）空气质量标准

biochemical oxygen demand 生物化学需氧量

food chain 食物链

（三）音译

音译就是根据英语单词的发音译成读音与原词大致相同的汉字。音译主要用于人名、地名、公司名及商标名等专有名词的翻译，例如 bit（比特）、joule（焦耳）、nylon（尼龙）、vaseline（凡士林）、morphine（吗啡）、El Nino（厄尔尼诺)、decibel（分贝）等。翻译这类词时需要译者耐心推敲，尽量遵守贴切、精炼的原则，努力达到音意兼顾、优雅易记的效果。

(四) 形译

形译指的是象形译，主要包括以下两种情况：①根据所指物体的形状进行翻译，即把原语词语中表示物体形状的首字母或首词译成译语中的形象词，其余部分意译，例如 U bolt（马蹄螺栓）、V-belt（三角带）、twist drill（麻花钻）；②根据词的形状进行翻译，即保留原词的首字母不译，其余部分意译，例如 S-turning（S形弯道）、A-bedplate（A形底座）、L-electron（L层电子）。

科技英语摘要写作（四）

摘要的主题句

论文摘要的第一句话通常被称为主题句（topic sentence），它的作用是开门见山地点名主题，概括出论文的主旨，乃至论文的主要内容、研究方法和研究发现。

主题句是表达整个论文摘要核心思想的句子，对摘要具有统括功能。但不是任何一个句子都可以做主题句。主题句通常有以下一些特点。论文摘要的主题句首先应是表达完整、充实的陈述句而不应该是一个词组或是小句。主题句要明确表达摘要的主旨，就不能太过笼统、空泛、抽象，相反，主题句一定要具体、确切、鲜明。

主题句时态、语态的选择也有着自己体例上的规约。在陈述"论文写作"的目的时，应该使用一般现在时，因为写论文是正在发生的事情；而在介绍具体的"研究目的"的时候，常常使用一般过去时，因为"研究目的"在实验伊始就已经确定了。上述两种情况多采用主动语态。然而当我们在介绍实验研究的具体内容时，通常是采取一般过去时和被动语态。

【例】1. The goal of the investigation is to experimentally demonstrate the feasibility of cogeneration with the nuclear heating reactor, raise its annual availability factor, and expand its application field.
2. Our objectives were to determine the variation in yield and other characteristics of corn grown in China.
3. An improvement in optical efficiency over contemporary external

reflector evacuated tube collectors was achieved by integrating the reflectors surface into the outer glass envelope.

能否有意识地、准确地写好主题句，这对摘要写作十分重要。没有写主题句的意识，不能地道地应用这些主题句，就会使得摘要写作的中心思想不突出，行文不够规范。因此，在平时的摘要写作练习中要注意强化这一意识，熟记上述提到的主题句句型表达，这对英语论文摘要的写作水平的提高会有很大的促进作用。

摘要译写示例

示例一

【摘要中文原文】

摘要：本文分析了啤酒生产中废水产生的环节、污染物及主要污染来源，并从好氧、厌氧生物处理两方面来考虑了废水治理工艺，提出了上流式厌氧污泥床（UASB）＋循环式活性污泥法（CASS）的组合工艺流程。可将废水COD由2,100 mg/L降至50～100 mg/L，BOD从1,100 mg/L降至20 mg/L以下，SS由300 mg/L降到70 mg/L以下，出水符合啤酒工业污染物排放标准（GB 19821—2005）。本设计工艺流程为：啤酒废水→格栅→集水池（污水提升泵房）→调节池→UASB反应器→CASS池→消毒池→出水。

【摘要原版译文】

Abstract: This paper analyzes the links of beer wastewater production, the pollutants and the major pollution sources, and proposes the Upflow Anaerobic Sludge Blanket (UASB) + loop activated Sludge (CASS) combination process in consideration of the wastewater treatment process from the aspects of aerobic and anaerobic biological treatment. The combination process can reduce the COD in wastewater from 2,100 mg/L to 50～100 mg/L, BOD from 1,100 mg/L to 20 mg/L or less, SS from 300 mg/L to 70 mg/L or less so that the effluent fits beer industry emission standards (GB 19821—2005). The technological process in the thesis is as following: beer wastewater → grilling → sump (sewage lift pump) → regulating pool → UASB reactor tank →

CASS pool → disinfection pool → the effluent.

【摘要修改译文】

Abstract: This paper studies the production processes of wastewater, the pollutants and the major pollution sources in beer production, and analyzes wastewater treatment methods from the perspective of aerobic and anaerobic biological treatment. The technological mix is put forward to combine the Upflow Anaerobic Sludge Blanket (UASB) with the Cyclic Activated Sludge System (CASS). The technological mix can reduce the COD of wastewater from 2,100 mg/L to 50–100 mg/L, BOD from 1,100 mg/L to 20 mg/L or below, SS from 300 mg/L to 70 mg/L or below so that the effluent fits beer industry emission standards (GB 19821—2005). The technological process in the study is as follows: beer wastewater → grilling → sump (sewage lift pump) → regulating pool → UASB reactor tank → CASS pool → disinfection pool → effluent.

【主要修改意见】

1. 原文第一句概括介绍了研究的目的和取得的结果，建议这两个内容分拆为两句，英文摘要不同于文章正文，应尽量使用简单句。同时对于提出的废水治理工艺应该用被动语态"The technological mix is put forward"来表达，更为客观。
2. 要注意专业术语表达的准确性，CASS 的全称不是"loop activated Sludge"，而应改为"Cyclic Activated Sludge System"。
3. 文中最后一句"as following"表达不对，要改为"as follows"。

示例二

【摘要中文原文】

摘要：设计了腐殖土填料藻浆厌氧发酵肥效反应器，进行为期 8 个月的臭味去除试验研究。动态对比试验结果表明，将藻浆上浮比控制于 75%，发酵温度控制于 20℃，pH 控制于 6.5~7.5 等条件下，腐殖土填料可大大提高藻浆厌氧发酵的速率，其藻浆厌氧发酵所需时间可缩短 5 d，厌氧发酵至 10 d 即可腐熟；在不降低藻浆厌氧发酵肥效的前提下，腐殖土填料对藻浆厌氧发酵臭味去除能力十分显著，腐熟阶段的臭味去除率超过 76%；温度变化对腐殖土填料的臭味去除能力有一定的影响，当温度从 30 ℃降至 15 ℃时，腐殖土填料对藻泥滤后液体的臭味去除率从 74.92% 升至 77.45%。综上，腐殖土填料具有显著的提升藻浆厌氧发酵速率和臭味去除能力，从而大大促进藻浆厌氧发酵作为有机肥推广应用的可行性。

【摘要原版译文】

　　Abstract: Humus soil reactor was designed to treat algae paste, and the trial removal of TON lasted 8 months. The results indicate that humus soil had significant effect on anaerobic fermentation speed under the conditions of floating ratio 75%, temperature 20℃ and pH 6.5–7.5. When humus soil was inoculated into reactor, anaerobic fermentation duration was shortened from 15d to 10d. With high fertilizer effectiveness, TON removal rate of anaerobic fermentation in humus soil reactor was over 76%. Temperature had effect on TON removal rate of anaerobic fermentation. TON removal rate of anaerobic fermentation was enhanced from 74.92% to 77.45% when temperature was reduced from 30℃ to 15℃. In conclusion, the feasibility of using algae paste anaerobic fermentation as organic fertilizer was greatly improved as humus soil has significant effect on anaerobic fermentation speed and TON removal rate of anaerobic fermentation.

【摘要修改译文】

　　Abstract: Humus soil and algae paste anaerobic fermentation reactor was designed to check fertilizer efficiency, and eight-month research was carried out to remove TON. The results indicated that humus soil could improve the ration of algae paste anaerobic fermentation on condition that the floating ratio was adjusted to 75%, temperature to 20℃ and pH to 6.5–7.5. When humus soil was inoculated into reactor, anaerobic fermentation duration was shortened from 15 d to 10 d, at which anaerobic fermentation could become thoroughly decomposed. With high fertilizer effectiveness, TON removal rate of anaerobic fermentation in humus soil reactor was over 76%. Temperature had effect on TON removal rate of anaerobic fermentation. TON removal rate of anaerobic fermentation was enhanced from 74.92% to 77.45% when temperature fell from 30℃ to 15℃. In conclusion, the feasibility of using algae paste anaerobic fermentation as organic fertilizer will be greatly improved as humus soil has significant effect on enhancing the anaerobic fermentation speed and TON removal rate of anaerobic fermentation.

【主要修改意见】

1. 第一句改为被动语态，因为科技论文的摘要一般用被动语态来表达，以避免提及研究者，使行文更客观。

2. 第二句中"had significant effect on",不能说明摘要中"可大大提高藻浆厌氧发酵的速率"的含义,所以改为"could improve the ration of"。
3. 第三句漏译了"厌氧发酵至10d即可腐熟",建议用定语从句来补充。
4. 最后一句得出结论,并对有机肥未来推广应用的前景进行展望,所以用一般将来时,"was greatly improved"改为"will be greatly improved"。

Unit 5　Sociology

Text A

Measuring Household and Living Arrangements of Older Persons Around the World: The United Nations Database on the Households and Living Arrangements of Older Persons 2019

　　Population ageing is occurring everywhere: nearly every country in the world is experiencing a substantial increase in the size and the proportion of the population aged 65 years or over. There were approximately 727 million persons aged 65 years or over in the world in 2020 and their number is projected to double to 1.5 billion in 2050. Globally, the share of the population aged 65 years or over is projected to increase from 9 per cent in 2020 to 16 per cent by 2050, so that one in six people in the world will be aged 65 years or over.

　　Population ageing is occurring along with broader social and economic changes that are taking place around the world. Decline in fertility, changes in patterns of marriage, cohabitation and divorce, increased levels of education among the younger generations, urbanization and migration in tandem with rapid economic development reshape the context in which older persons live, including the size and composition of their households and their living arrangements. Many of these changes raise concerns about a possible weakening of the traditional family, which, historically, has been the foundation of economic security and social support for older persons in many parts of the world. In countries for which historical data are available (primarily those located in Western Europe and Northern America), intergenerational co-residence has declined dramatically and most older persons currently live either in single-person households or in households consisting of a couple only or a couple and their unmarried children. Available data indicate that many countries in less developed regions are also experiencing a slow shift in family and household composition away from multi-generational households towards the nuclear family households that are more prevalent in Western Europe and in the United States.

While family and household structures change rather slowly, external shocks, such as economic or health crises, often call on families to quickly react to provide the support needed to their kin. The impact of the 2008 financial crisis and austerity policies in Greece, Italy and Spain led adult children with families to move back in with their older parents. In the United States, social and economic crises such as the crack and opioid epidemics, mass incarceration, and child welfare policies that separate children from their parents have contributed to a rising number of skip generation households, especially among African American families. In some sub-Saharan African, Latin American and Caribbean countries, increasing numbers of older people are becoming heads of households and primary careers for family members and children, whose parents are absent as a result of the HIV/AIDS epidemic or labor migration. Armed conflict, social and ethnic tensions along with public health care emergencies, economic crises, as well as environmental disasters impact directly and indirectly the sustainability of social, community and family support networks of older persons in many parts of the world. Each of these transformations is reshaping the contexts in which older persons live.

Living arrangements of older persons are an important determinant of their physical well-being as well as their morbidity and mortality. The living arrangements of older persons have been associated with their economic well-being, physical and psychosocial health and life satisfaction. Household size and living arrangements of older persons can also have important macroeconomic implications by influencing the demand for housing, social services, energy, fuel, water and other resources.

Globally, the majority of older persons in Northern America and Europe live independently, i.e., alone or with their spouse only. Research has pointed out that most of the older persons in these countries prefer to live independently as long as their health is good enough to do so. In this context, living independently is a matter of preference as privacy is usually considered a normal good. However, living independently does not necessarily indicate an absence of family support. Often, older parents and adult children maintain households nearby and help each other by exchanging financial support, informal care, and other forms of assistance even when they live apart. In the United States, for example, nearly a quarter of older parents lived within 1 mile of a child, and 60 per cent had at least one child located within 10 miles, according to data from the 1987-1988 National Survey of Families and Households. Data from the U.S. Health and Retirement Study (HRS) showed that unmarried mothers and their adult children tended to live in close proximity, with about one third residing within 10 miles of each other. As some persons living alone are never married and childless, they are more likely to rely on other relatives (siblings and other kin) as well as non-kin (friends, neighbors) for contact and support.

Likewise, older persons in less developed countries who do not co-reside with their adult children also tend live near their children's households. According to the China Health and Retirement Longitudinal Study (CHARLS) baseline survey, among those individuals aged 60 or over with at least one adult child, 41 per cent co-reside with their adult child, 34 per cent live in the same immediate neighborhood as the adult child and 14 per cent live in the same county. Only 5 per cent do not have an adult child at least within the same county.

In most of the less developed countries, the majority of older persons live with their children or with extended family members. In many of these countries, the absence of comprehensive social protection programs and declining labor market prospects of adult children, co-residence of older parents with children is an important element of the flow of financial, emotional and care support between family members.

The United Nations Database on the Households and Living Arrangements of Older Persons 2019 is the only dataset that provides harmonized and comparable data on patterns and trends in the household size and composition and the living arrangements of older persons at the global level, across regions and countries, and over time. Although the database cannot provide the networks, pathways and direction of support between older persons and their kin, this database is a unique data sources of estimates from 155 countries or areas, representing more than 97 per cent of persons aged 60 or over globally, with reference dates ranging from 1960 to 2018. It provides estimates of the global patterns and trends of household size, composition and the living arrangements of the older persons. The dataset distinguishes co-residence according to the age of children, i.e., older persons living with children under age 20 and those living with adult children 20 years or over, an important distinction to better understand the co-residence with children as part of the life course of older persons. Because most of the data sources accessed relied on information about households, older persons residing in institutions such as nursing facilities, prisons, religious institutions or dormitories are not represented in the data. The estimates should thus be interpreted as referring to the household population only. The analyses in this report refers to older persons aged 65 or over and includes one recent observation for 153 of the countries with reference dates ranging from 2000 to 2018, representing 97 per cent of persons aged 65 or over globally.

(1,191 words)

New Words

1. approximately [əˈprɔksimətli]　*adv.* 大约
2. fertility [fəˈtiləti]　*n.* 生育
3. cohabitation [ˌkəuˌhæbiˈteiʃn]　*n.* 同居
4. urbanization [ˌəːbənaiˈzeiʃn]　*n.* 城市化
5. migration [maiˈgreiʃn]　*n.* 移民
6. reshape [ˌriːˈʃeip]　*vt.* 改变；重塑
7. context [ˈkɔntekst]　*n.* 背景
8. composition [ˌkɔmpəˈziʃn]　*n.* 构成
9. household [ˈhaushəuld]　*n.* 家庭；家务
10. primarily [praiˈmerəli]　*adv.* 主要地
11. intergenerational [ˌintədʒenəˈreiʃənl]　*adj.* 两代之间的；代际的
12. indicate [ˈindikeit]　*vt.* 表明
13. multi-generational [ˌmʌltidʒenərˈeiʃənəl]　*adj.* 多代同堂的
14. prevalent [ˈprevələnt]　*adj.* 盛行的；普遍存在的
15. external [ikˈstəːnl]　*adj.* 外来的；外在的
16. kin [kin]　*n.* 亲戚

17. austerity [ɔːˈsterəti] *n.* 经济紧缩
18. crack [kræk] *n.* 强效纯可卡因
19. opioid [əʊˈpiːəʊid] *adj.* 与阿片类药物有关的
20. epidemic [ˌepiˈdemik] *n.* 传染病；流行病
21. mass [mæs] *adj.* 人数众多的
22. incarceration [inˌkɑːsəˈreiʃn] *n.* 监禁；禁闭
23. sub-Saharan [ˌsʌbsəˈhɑːrən] *adj.* 撒哈拉以南的
24. Caribbean [ˌkæriˈbiːən] *adj.* 加勒比海的；加勒比海人的
25. ethnic [ˈeθnik] *adj.* 种族的
26. tension [ˈtenʃn] *n.* 冲突
27. emergency [iˈməːdʒənsi] *n.* 突发事件；紧急情况
28. transformation [ˌtrænsfəˈmeiʃn] *n.* 转型
29. determinant [diˈtəːminənt] *n.* 决定因素
30. morbidity [mɔːˈbidəti] *n.* 发病率
31. mortality [mɔːˈtæləti] *n.* 死亡率
32. psychosocial [saikəʊˈsəʊʃəl] *adj.* 社会心理的
33. macroeconomic [ˌmækrəʊˌiːkəˈnɔmik] *adj.* 宏观经济的
34. implication [ˌimpliˈkeiʃn] *n.* 意义
35. majority [məˈdʒɔrəti] *n.* 大多数
36. spouse [spaus] *n.* 配偶
37. privacy [ˈprivəsi] *n.* 隐私
38. assistance [əˈsistəns] *n.* 援助
39. proximity [prɔkˈsiməti] *n.* 邻近
40. longitudinal [lɔŋgiˈtjuːdinl] *adj.* 纵向的
41. baseline [ˈbeislain] *n.* 基线
42. comprehensive [ˌkɔmpriˈhensiv] *adj.* 综合性的；全面的
43. database [ˈdeitəbeis] *n.* 数据库
44. dataset [ˈdeitəset] *n.* 数据源
45. harmonize [ˈhɑːmənaiz] *vt.* 使一致
46. comparable [ˈkɔmpərəbl] *adj.* 可比的
47. pathway [ˈpɑːθwei] *n.* 途径
48. estimate [ˈestimeət] *n.* 估计
49. distinguish [diˈstiŋgwiʃ] *vt.* 辨别出
50. distinction [diˈstiŋkʃn] *n.* 区别

51. course [kɔːs]   *n.* 历程
52. institution [ˌɪnstɪˈtjuːʃn]   *n.* 机构

Useful Expressions

1. be projected to   预计
2. in tandem with   同，同…合作
3. multi-generational household   几世同堂的家庭
4. nuclear family   小家庭（只包括父母和子女的家庭）
5. mass incarceration   大规模监禁
6. child welfare policy   儿童福利政策
7. skip generation household   隔代家庭
8. public health care emergency   突发公共卫生事件
9. the majority of   大多数
10. extended family   大家庭
11. life course   生命历程

Notes

1. 本文选自联合国（United Nations，UN）官网（https://www.un.org），https://www.un.org/development/desa/pd/sites/www.un.org.development.desa.pd/files/desa-pd-technicalpaper-living_arrangements_of_older_persons_2019.pdf。

2. 美国健康与退休调查（U.S. Health and Retirement Study，HRS），是由密歇根大学主持的研究项目，得到了国家老龄化研究所和社会保障管理局的大力支持。此研究针对 20 000 名 50 岁以上的美国人，每两年进行一次调查，是世界最大的追踪调查项目之一。

3. 中国健康与养老追踪调查（China Health and Retirement Longitudinal Study，CHARLS），是由北京大学国家发展研究院主持、中国社会科学调查中心执行的一项大型跨学科调查项目，是国家自然科学基金委员会资助的重大项目。项目负责人是北京大学国家发展研究院经济学教授。此研究采取基线样本追踪访问的研究模式，随着多年调查的持续进行，能够记录同一批经过科学方法抽象的、具有我国居民普遍代表性的受访者的发展变化情况，访问成果既有即时的科研价值，又有多维度跨学科的历史价值。调查实施越持久，追访越

深入，其学术价值越高。

4. 基线调查（baseline survey），亦称基础调查。在不同政策的调查中使用不同的方法，总地来说就是线索收集、问卷调配、调查内容不同，方法也有所不同。例如，出生缺陷干预工程的基线调查，其核心是出生缺陷发生现状的调查。除此之外还包括：出生缺陷发生原因的调查、群众对出生缺陷认识程度的调查、服务机构能力的调查（如资料收集能力、技术服务能力、管理及服务的规范能力等）、服务机构现状的调查（也称作基础性调查或干预前调查）。它的性质和计划生育工作中的摸底调查是相似的。

Exercises

Part Ⅰ Vocabulary and Structure

Section A   Match each word with its Chinese equivalent.

1. morbidity   A. 生育
2. epidemic   B. 家庭
3. kin   C. 亲戚
4. household   D. 经济紧缩
5. database   E. 传染病
6. mortality   F. 监禁
7. ethnic   G. 种族的
8. incarceration   H. 发病率
9. austerity   I. 死亡率
10. fertility   J. 数据库

Section B   Fill in the blanks with the words or expressions given below. Change the form where necessary.

| privacy | indicate | majority | tension | distinguish |
| determinant | urbanization | reshape | context | prevalent |

1. A survey of retired people has _____ that most are independent and enjoying life.
2. Despite the fact that the disease is so _____, treatment is still far from

satisfactory.
3. Asset (资产) allocation (配置) ends up being the overwhelmingly important _____ of the results.
4. Microsoft is committed (使致力于) to openness, innovation and the protection of _____ on the internet.
5. Research suggests that babies learn to see by _____ between areas of light and dark.
6. Female workers constitute the _____ of the labor force.
7. We are doing this work in the _____ of reforms in the economic, social and cultural spheres.
8. The film explored the _____ between public duty and personal affections.
9. If they succeed, they will have _____ the political and economic map of the world.
10. _____ is the process of creating cities or towns in country.

Part Ⅱ Translation

Section A Translate the following sentences into Chinese.

1. Population ageing is occurring everywhere: nearly every country in the world is experiencing a substantial increase in the size and the proportion of the population aged 65 years or over.
2. While family and household structures change rather slowly, external shocks, such as economic or health crises, often call on families to quickly react to provide the support needed to their kin.
3. The living arrangements of older persons have been associated with their economic well-being, physical and psychosocial health and life satisfaction.
4. As some persons living alone are never married and childless, they are more likely to rely on other relatives (siblings and other kin) as well as non-kin (friends, neighbors) for contact and support.
5. Because most of the data sources accessed relied on information about households, older persons residing in institutions such as nursing facilities, prisons, religious institutions or dormitories are not represented in the data.

Section B Translate the following sentences into English.

1. 他们的研究表明该国婴儿的死亡率已降至历史最低水平。
2. 不同种族背景的人曾共同生活在那个地区。
3. 根据他们的调查，半数以上的家庭申报收入超过 35 000 英镑。
4. 关于社会学的作者索引(indexes)可在图书馆的数据库中找到。
5. 近 50 年来，公众对婚姻的态度已经发生改变。

Text B

Introduction to Sociology

Concerts, sports games, and political *rallies*（集会）can have very large crowds. When you attend one of these events, you may know only the people you came with. Yet you may experience a feeling of connection to the group. You are one of the crowd. You cheer and applaud when everyone else does. You *boo*（发出嘘声）and yell alongside them. You move out of the way when someone needs to get by, and you say "excuse me" when you need to leave. You know how to behave in this kind of crowd.

It can be a very different experience if you are travelling in a foreign country and find yourself in a crowd moving down the street. You may have trouble figuring out what is happening. Is the crowd just the usual morning rush, or is it a political protest of some kind? Perhaps there was some sort of accident or disaster. Is it safe in this crowd, or should you try to extract yourself? How can you find out what is going on? Although you are in it, you may not feel like you are part of this crowd. You may not know what to do or how to behave.

Even within one type of crowd, different groups exist and different behaviours are on display. At a rock concert, for example, some may enjoy singing along, others may prefer to sit and observe, while still others may join in a *mosh pit*（舞池）or try crowd surfing. On February 28, 2010, Sidney Crosby scored the winning goal against the United States team in the gold medal hockey game at the Vancouver Winter Olympics. Two hundred thousand **jubilant** people filled the streets of downtown Vancouver to celebrate and cap off two weeks of uncharacteristically *vibrant*（充满生机的）, joyful street life in Vancouver. Just over a year later, on June 15, 2011, the Vancouver Canucks lost the seventh hockey game of the Stanley Cup finals against the Boston Bruins. One hundred thousand people had been watching the game on outdoor screens. Eventually 155,000 people filled the downtown streets. *Rioting*（暴乱）and *looting*（抢劫）led to hundreds of injuries, burnt cars, trashed storefronts and property damage totaling an estimated $4.2 million. Why was the crowd response to the two events so different?

A key insight of sociology is that the simple fact of being in a group changes your behaviour. The group is a phenomenon that is more than the sum of its parts. Why do we feel and act differently in different types of social situations? Why might people of a single group exhibit different behaviours in the same situation? Why might people acting similarly not feel connected to others exhibiting the same behaviour? These are some of the many questions sociologists ask as they study people and societies.

What Is Sociology?

A dictionary defines sociology as the systematic study of society and social interaction. The word "sociology" is derived from the Latin word *socius* (companion) and the Greek word *logos* (speech or reason), which together mean "reasoned speech about companionship". How can the experience of companionship or togetherness be put into words or explained? While this is a starting point for the discipline, sociology is actually much more complex. It uses many different methods to study a wide range of subject matter and to apply these studies to the real world.

The sociologist Dorothy Smith defines *the social* as the "ongoing concerting and coordinating of individuals' activities". Sociology is the systematic study of all those aspects of life **designated** by the adjective "social". These aspects of social life never simply occur; they are organized processes. They can be the *briefest*（简洁的）of everyday interactions—moving to the right to let someone pass on a busy sidewalk, for example—or the largest and most enduring interactions—such as the billions of daily exchanges. If there are at least two people involved, even in the *seclusion*（与世隔绝的地方）of one's mind, then there is a social interaction that entails the "ongoing *concerting*（协调）and coordinating of activities". Why does the person move to the right on the sidewalk? What collective process lead to the decision that moving to the right rather than the left is normal?

What Are Society and Culture?

Sociologists study all aspects and levels of society. A society is a group of people whose members interact, reside in a definable area, and share a culture. A culture includes the group's shared practices, values, beliefs, norms and *artifacts* (人工产品). One sociologist might analyze video of people from different societies as they carry on everyday conversations to study the rules of polite conversation from different world cultures. Another sociologist might interview a representative sample of people to see how email and instant messaging have changed the way organizations are run. Yet another sociologist might study how migration determined the way in which language spread and changed over time. A fourth sociologist might study the history of international agencies like the United Nations or the International Monetary Fund to examine how the globe became divided into a First World and a Third World after the end of the colonial era.

These examples illustrate the ways society and culture can be studied at different levels of analysis, from the detailed study of face-to-face interactions to the examination of large-scale historical processes affecting entire civilizations. It is common to divide these levels of analysis into different gradations based on the scale of interaction involved. Sociologists break the study of society down into four separate levels of analysis: micro, meso, macro, and global. The basic distinction, however, is between micro-sociology and macro-sociology.

The study of cultural rules of politeness in conversation is an example of micro-sociology. At the micro-level of analysis, the focus is on the social *dynamics* (动态) of intimate, face-to-face interactions. Research is conducted with a specific set of individuals such as conversational partners, family members, work associates, or friendship groups. In the conversation study example, sociologists might try to determine how people from different cultures interpret each other's behaviour to see how different rules of politeness lead to misunderstandings. If the same misunderstandings occur consistently in a number of different interactions, the sociologists may be able to propose some generalizations about rules of politeness that would be helpful in reducing tensions in mixed-group dynamics (e.g., during staff meetings or international negotiations).

Macro-sociology focuses on the *properties* (特性) of large-scale, society-wide social interactions: the dynamics of institutions, classes, or whole societies. The example above of the influence of migration on changing patterns of language usage is a macro-level phenomenon because it refers to structures or processes of social interaction that occur outside or beyond the intimate circle of individual social *acquaintances* (泛泛之交). These include the economic and other circumstances that lead to migration; the educational, media, and other communication structures that help or hinder the spread of speech patterns; the class, racial, or ethnic divisions that create different *slangs* (俚语) or cultures of language use; the relative isolation or integration of different communities within a population; and so on. Other examples of macro-level research include examining why women are far less likely than men to reach positions of power in society or why *fundamentalist* (原教旨主义者) Christian religious movements play a more prominent role in American politics than they do in Canadian politics. In each case, the site of the analysis shifts away from the nuances and detail of micro-level interpersonal life to the broader, macro-level systematic patterns that structure social change and social cohesion in society.

The Sociological Imagination

Obesity, for example, has been increasingly recognized as a growing problem for both children and adults in North America. Michael Pollan cites statistics that three out of five Americans are overweight and one out of five is obese. In Canada in 2012, just under one in five adults (18.4 percent) were obese, up from 16 percent of men and 14.5 percent of women in 2003 (Statistics Canada 2013). Obesity is therefore not simply a private trouble concerning the medical issues, dietary practices, or exercise habits of specific individuals. It is a widely shared social issue that puts people at risk for chronic diseases like hypertension, diabetes, and *cardiovascular* (心血管的) disease. It also creates significant social costs for the medical system.

(1,373 words)

Comprehension of the Text

Choose the best answer to each of the following questions.

1. Which of the following topics is not mentioned in the passage?
   A. The definition of sociology.
   B. The importance of sociology.
   C. The different levels of analysis in sociology.
   D. The research methods used in sociology.
2. During the concerts or sports games why do people cheer and applaud alongside the others from the sociological perspective?
   A. Because it is a kind of social convention.
   B. Because people share a feeling of connection to the group.
   C. Because it can show their good manner.
   D. Because they enjoyed the program.
3. What does the word "jubilant" (Line 7, Para. 3) probably mean?
   A. Depressed.   B. Anxious.
   C. Joyful.   D. Disappointed.
4. The study focus of sociology is on _____ .
   A. the process of human evolution
   B. the civilization of mankind society
   C. the relationship between human and societies
   D. the interaction between human and nature
5. Which of the following words has the similar meaning with "designated" (Line 3, Para. 6)?
   A. Specified.   B. Clarified.
   C. Simplified.   D. Nominated.
6. When it comes to society and culture, which of the following statements is NOT correct?
   A. Sociologists might study rule of polite conversation.
   B. Sociologists might examine how emails have changed the way organization worked.
   C. Sociologists might study how migration affect the changes of language.
   D. Sociologists might study the sound system of a specific language.
7. The ways society and culture can be divided into various gradations rely on

_____.

    A. scale of interaction   B. effectiveness of interaction

    C. frequency of interaction   D. purpose of interaction

8. In terms of micro-level of analysis, in the conversation study example, which is NOT the focus of sociologists?

    A. To determine how people explain each other's behaviour from different cultures.

    B. To analyze the reasons of culture shock.

    C. To find out how different rules of politeness result in misunderstanding from different cultures.

    D. To propose some generalizations about rules of politeness.

9. The research about why women are far less likely than men to reach position of power in society belongs to _____ level of analysis.

    A. micro-sociology   B. meso-sociology

    C. macro-sociology   D. global-sociology

10. According to the passage, obesity is both a medical issue and a social issue which can result in some chronic diseases except _____.

    A. high blood pressure   B. diabetes

    C. cardiovascular disease   D. stomach trouble

科技文体翻译技巧(五)

数量的译法

数量的描述在科技英语论文写作中占有不可或缺的地位。当代科学的发展需要各种类型数据的支撑，英汉关于数量的表述有所不同。因此在数量词翻译的过程中，需要仔细推敲，才能够避免"失之毫厘，谬以千里"的错误。

一、概括数量

从整体数量把握，进行概括性的描述。"数量"在英语中通常译成名词，表示可数名词的"数量"译成 number，不可数名词的"数量"译成 amount、quantity 或 volume，常用形容词 large、great、considerable、sufficient、moderate、small 等修饰。

【例】1. As we know, the number of chromosomes in animal cells varies from species to species.

据我们所知，动物细胞内染色体的数目因物种的差异而有所不同。

2. When the animal exhales, large amounts of carbon dioxide are expelled.

当动物呼气时，排出大量的二氧化碳。

二、含量

在某物质中某成分的含量是多少，其表达方法，通常利用动词、动词词组或名词词组，常用的动词或动词短语有 contain、comprise、constitute、include、form、consist of、make up、be composed of 等，名词短语有 the component(s) of、the content(s) of、the composition(s) of、the constituent(s) of 等。

【例】1. Air that is inhaled into the lungs contains large amounts of oxygen.

吸入肺部的空气含有大量的氧气。

2. Sunflower margarine has the same fat content as butter.

向日葵所制人造黄油的脂肪含量与黄油脂肪含量相同。

三、尺寸

尺寸是指物体的长度、宽度、高度、容积和体积。尺寸的表达方式有以下3种。

（一）物＋be＋数量词＋形容词

【例】1. The horse in the region is only 80 centimeters high.

这个地区的马只有 80 cm（厘米）高。

2. The water is 3 feet deep.

水有 3 ft（英尺）深。

（二）物＋have＋尺寸抽象词＋数量词

【例】1. The pipe has a length of 1.5 meters.

管子长 1.5 m（米）。

2. The egg had a shell thickness of 0.14 mm and the newly-hatched chick weighed 9 grams.

那只蛋的壳厚 0.14 mm（毫米），新孵出的雏鸟重 9 g（克）。

（三）尺寸抽象词＋物＋be＋数量词

【例】1. The height of this tree is about 20 meters.

这棵树高约 20 m（米）。

2. The width of the river is 10 meters.

 河宽 10 m（米）。

四、数量的修饰

科技英语数量的表述可与适当的修饰语相搭配，从而使表达更加准确。

（一）数量词 exact 和 exactly

【例】Each corner had a guard tower, each of which was exactly 10 meters in height.

每个角落都有一座警戒塔，每座警戒塔正好 10 m（米）高。

（二）平均数 average、mean 和 on average

【例】The average price of agricultural products rose by just 2.2%.

农产品的平均价格只上涨了 2.2%。

（三）大约数 about、approximate、approximately、estimated 和 close to

【例】Approximately $150 million is to be spent on improvements in farmers' life.

大约 1.5 亿美元将用于改善农民的生活。

（四）超过数 over、above 和 more than

【例】The temperature crept up to above 40 degrees.

温度攀升至 40 度出头。

（五）不足数 under、below 和 less than

【例】Motorways actually cover less than 0.1 percent of the countryside.

高速公路的实际覆盖面积还不到农村地区的 0.1%。

（六）范围数 from... to... 和 between... and...

【例】The layer of gases is about from 200 to 250 miles in thickness.

大气层厚为 200~250 英里。

（七）另加数 more 和 further

【例】Twenty more grams of salt was added to the water.

给水又添加了 20 g（克）的盐。

（八）单位数 every 和 each

【例】Each of the machines is perhaps five feet in diameter.

这些机器每台直径大约 5 ft（英尺）。

（九）强调数 only、at most、at least 和 as... as

【例】The speed exceeds the limit by a factor of two at least.

该速度至少超过限定速度一倍。

（十）总数 total、amount to、come to 和 in number

【例】The total number of cows dying from bovine spongiform encephalopathy (BSE) is estimated to reach 20,000.

患疯牛病死亡的牛的总数预计达 2 万头。

（十一）间隔数 interval、alternate、every 和 no less than

【例】The temperature was recorded every five hours.

每 5 h（小时）记录一次温度。

（十二）最高数和最低数 top、maximum 和 minimum

【例】The maximum weight of the instrument is 6 pounds.

这台仪器的最大重量是 6 lb（磅）。

五、倍数的增减

（一）倍数的增加

英语涉及倍数的增加，通常指增加到的程度，英译汉的过程中，表示净加的倍数时，应将原文的倍数减一，汉译英时，应将原来的倍数加一。

1. 动词＋by n times /n-fold

用得较多的动词有 increase、grow 和 rise，有时用 raise、soar、enlarge、extend、expand、exceed、surpass、multiply、go up 等。

【例】The stocking rate can be increased by three times after the pastures have been improved.

草场改良后，载畜量可以提高 2 倍。

2. n times＋as＋形容词或副词＋as

【例】The output of fertilizers of this region last year was 11 times as much as that in 2010.

去年，这个地区的化肥产量比 2010 年增加 10 倍。

3. be＋n times＋that/over、be＋n times＋名词＋of 和 be＋up＋n times/n-fold

【例】The yield of grain for 2014 was five times over that for 1989.

2014 年的粮食产量比 1989 年多 4 倍。

4. n times＋比较级＋than

【例】The new type of firebrick lasts at least three times longer than other firebricks.

这种新型耐火砖比其他耐火砖使用寿命至少长 2 倍。

5. n times/n-fold＋名词

【例】The number of farms in this area has an increase of six fold from 2000 to 2010.

这个地区的农场数目在2000—2010年增长5倍。

（二）倍数的减少

英语中对倍数减少的表述所使用的数词同倍数增加的数词保持一致，英语的"减少 n 倍"相当于汉语的"减少到 $1/n$"或"减少了 $n-1/n$"。

1. 动词＋by n (times) /n-fold

这样的动词有 reduce、decrease、drop、decline、lesson、shorten、cut down 等。

【例】Technical innovation in agriculture reduced the cost of production by four times.

农业的技术革命使生产成本降低了3/4。

2. n times＋名词

这类名词有 reduction、decrease、drop、decline 等。

【例】In the last fiscal year the expenditure on capital construction had a five times decline compared with the fiscal year before last.

在上个财政年度，基本建设开支比上个财政年度减少80%。

3. n times＋as＋形容词或副词＋as

【例】The convalescence of rheumatic patients treated by acupuncture is generally half as short as the convalescence of those treated by other therapy.

用针灸治疗的风湿症患者的恢复期，一般来说比用其他疗法短1/2。

4. n times＋比较级＋than

【例】With an increasing number of households using gas, coal consumption as fuel in this city is now three times less than ten years ago.

由于越来越多的住户使用煤气，这座城市燃料煤的消耗比十年前减少了2/3。

科技英语摘要写作（五）

摘要的展开句

科技文体摘要的展开句是指根据摘要的主题句，对文章主体的内容进行概括性的陈述，可涉及研究的内容、方法及结果，框架部分通常由3～5句构成。

有两点注意事项：①语态上，所使用的语态以主动语态为主，被动语态为辅。②时态上，实验研究类摘要多数句子使用一般过去时，表示研究是业已完成的动作，如实验的目的、结果等；少数句子使用一般现在时，表达研究中普遍接受的事实，如实验的方法和结论；文献综述类摘要多数句子使用一般现在时，少数句子使用过去时，背景也可应用一般现在时。

摘要的展开句涉及多种语义关系的表达句型，包括比较、对比、认同、分类等。

【常用句型】

1. Comparison

... be similar to...

... be like/resembles...

2. Contrast

... be different from...

... contrast with...

3. Agreement/Disagreement with the author/source

... correctly notes...

... rightly observes...

However, it remains unclear whether...

... doesn't support one's conclusion that...

4. Classification

... can be divided /classified into... （and... ）

... and... be categories/divisions of...

5. Assertion

It can be claimed/assumed that...

It seems certain/doubtful that...

根据摘要展开句的内容总结归纳句型，可以涉及研究的目的、方法及结果。

【常用句型】

1. Aim

The present study was to....

The aim/purpose of the study/research was to....

2. Method

... were/was used/applied to do....

The data were evaluated using....

3. Result

We investigated/found....

Our result(s)/data showed/revealed/demonstrated that....

摘要译写示例

示例一

【摘要中文原文】

摘要：线粒体甘油-3-磷酸酰基转移酶基因（mitochondrial glycerol-3-phosphate acyltransferase，GPAM）的主要作用是通过催化三酰基甘油和磷脂生物合成过程，从而促进动物机体甘油三酯的生成，并通过中心和外周调节作用，对哺乳动物脂肪沉积、能量消耗、胴体重以及整体的代谢进行调控。本研究设计特定引物对牛 GPAM 基因编码区进行 PCR 扩增并进行了克隆及测序，并假定牛 GPAM 基因编码序列发生 E1-76-C>T 点突变，通过生物信息学的方法，对牛 GPAM 基因突变前后编码产物的理化性质、结构与功能进行了对比分析。通过生物信息学对比分析得出，牛 GPAM 基因单碱基的改变可以影响氨基酸的编码，在一级结构、二级结构和三级结构方面存在显著差异，为 GPAM 基因结构与功能的研究奠定了基础。

【摘要原版译文】

Abstract：The main function of mitochondrial glycerol-3-phosphate acyltransferase (GPAM) gene is through catalytic triacylglycerol phospholipid biosynthesis, thus promote the animal's body triglyceride, and through the center and peripheral regulate the energy consumption, carcass weight and fat deposition for mammals, the overall metabolic regulation. In this study, the coding region of the bovine GPAM gene was amplified via specific primers by PCR method, and the sequence was cloned and sequenced. Bioinformatics methods were taken for a comparative analysis of physical and chemical properties, structure and function between mutant type and normal type, assuming that bovine GPAM gene coding sequence with E1-76-C>T mutation site. Through the comparison of bioinformatics analysis, the bovine GPAM gene single nucleotide change can affect the amino acid coding, lead to there are significant differences in primary structure, secondary structure and three levels of protein. The research lays a foundation for the studies on the structure and function of bovine GPAM gene.

【摘要修改译文】

Abstract：The main function of mitochondrial glycerol-3-phosphate acyltransferase (GPAM) gene is to promote production of triglyceride in animal body through catalytic triacylglycerol phospholipid biosynthesis, and to regulate fat deposition, energy consumption, carcass weight and overall metabolism of mammals through central and peripheral regulating action. In this study, coding region of bovine GPAM gene was amplified via specific primers by PCR, and was cloned and sequenced. Bioinformatics methods were used to comparatively analyze physical and chemical properties, structure and function between mutant type and normal type, assuming that mutation occurred with E1-76-C>T site of bovine GPAM gene coding sequence. Comparative analysis of bioinformatics showed that, bovine GPAM gene single base change can affect amino acid coding, and there are significant differences in primary structure, secondary structure and tertiary structure of protein. The research lays a foundation for the studies on structure and function of bovine GPAM gene.

【主要修改意见】

1. 当实验对象的功能或研究目的做主语时，谓语应使用：系动词 be＋不定式短语 to do，如文中 "The main function of... is through..., thus promote the animal's body triglyceride" 应修改为 "The main function of... is to promote the animal's body triglyceride."

2. 应认真分析汉语长句的核心结构，避免出现中式英语的现象，原文汉语第一句的核心结构为"主要作用是促进动物机体甘油三酯的生成并调控哺乳动物脂肪沉积等"。

3. 注意主语与谓语的合理搭配，如 methods 做主语，谓语动词应使用 be used，而不是 be taken。

4. 一个动词的动名词同其名词形式相比，具有较强的动作趋向性，如原文汉语表述为"对…进行对比分析"，因此翻译为"comparatively analyzing"比"a comparative analysis of..."恰当。

5. 翻译时，应确保英文句子核心结构的完整性，防止发生缺少谓语的情况，如 "assuming that bovine GPAM gene coding sequence with E1-76-C>T mutation site" 中 that 从句没有谓语部分。

6. 注意定冠词 the 的用法，当文中首次提到某个话题时，可以不用特指，如结尾句中 "the structure and function of bovine GPAM gene" 中的 the 可以删除。

示例二

【摘要中文原文】

摘要：为发掘与鸡免疫性状相关的分子标记，本试验以大骨鸡为试验材料，采用半定量 RT-PCR 法对 *MYCN* 基因不同组织 mRNA 表达特性进行分析，并用 PCR-SSCP 方法对 *MYCN* 基因第二外显子的序列多态性进行检测，通过关联分析以期望发现与鸡免疫性状相关的单核苷酸变异。结果表明，鸡 *MYCN* 基因在心脏、肝脏、脾脏、肺、肾脏、卵巢、大脑和胸腺组织中均有表达，但其表达水平存在一定差异，其中，在脾脏和胸腺组织中的 mRNA 表达水平最高（$P<0.05$），在肺脏、肾脏、卵巢和大脑中的表达水平次之，在心脏和肝脏中的 mRNA 表达水平最低（$P<0.05$）。在 *MYCN* 基因第二外显子上检测到 C2178T 的转换，该多态性位点产生了 3 种基因型：AA、BB 和 AB，基因型频率分别为 0.358、0.058 和 0.584，等位基因 A 和 B 的频率则分别为 0.65 和 0.35。对不同基因型的 120 只大骨鸡进行免疫性状关联分析表明，BB 基因型个体的碱性磷酸酶含量显著高于 AA 型和 AB 型（$P<0.05$）。且 BB 基因型个体的红细胞含量和红细胞压积均显著高于 AA 型（$P<0.05$），其他免疫性状间的差异不显著（$P<0.05$）。此研究结果为进一步筛查并验证 *MYCN* 基因可否作为抗病候选基因提供参考依据。

【摘要原版译文】

Abstract：To explore the molecular markers associated with chicken immune traits, the experiment with Dagu chicken as test material, semi-quantitative RT-PCR method was used to analyze the expression of *MYCN* gene mRNA of different tissues, and PCR-SSCP method was used to detect the genetic polymorphisms of *MYCN* gene, and thus be analyzed with immune traits associated. The results showed that：*MYCN* gene is expressed in heart, liver, spleen, lung, kidney, ovary, brain and thymus of chicken. Among them, the highest expression level of mRNA is in the spleen and thymus, followed by lung, heart, liver, ovarian. This shows that chicken *MYCN* mRNA in vivo tissue widely distributed, but the expression level is uneven. The results of PCR-SSCP showed that a variation C2178T was detected on conversion exon Ⅱ of *MYCN* gene, three genotypes were found：AA, BB, AB. Genotype frequencies：0.358, 0.058, 0.584, allele frequencies：A0.65, B0.35. Different enzyme genotype and biochemical blood analysis showed an association：alkaline phosphatase levels of BB genotype were significantly higher than AA and AB ($P<0.05$); other blood biochemical enzyme content was not significant ($P>0.05$); different genotypes associated with some immune parameters showed that: levels of red blood cells and hematocrit of BB genotype were significantly higher than AA genotype ($P<0.05$); other immune parameters was not significant ($P<0.05$). It provided a reference data for us to further confirm *MYCN* gene act as an anti-disease candidate gene.

【摘要修改译文】

　　Abstract: To explore molecular Markers correlated with chicken immune traits, having Dagu chicken as test materials, mRNA expression characteristics of different tissues of *MYCN* gene were analyzed by semi-quantitative RT-PCR. Sequence polymorphisms of *MYCN* gene exon Ⅱ were detected by PCR-SSCP. Correlation analysis was used to study nucleotide variation correlated with chicken immune traits. Among them, the highest expression level of mRNA is in spleen and thymus ($P<0.05$), followed by lung, kidney, ovary and brain and the lowest is in heart and liver ($P<0.05$). C2178T transfer was detected on conversion exon Ⅱ of *MYCN* gene, and three genotypes, AA, BB and AB, were found on this polymorphic site. Genotype frequencies of AA, BB and AB are 0.358, 0.058 and 0.584 respectively, and allele frequencies of A and B are 0.65 and 0.35 respectively. Immune traits correlation analysis of 120 Dagu chicken of different enzyme genotypes showed that alkaline phosphatase levels of BB genotype are significantly higher than those of AA and AB ($P<0.05$); levels of red blood cells and hematocrit of BB genotype are significantly higher than those of AA genotype ($P<0.05$); differences among other immune traits are not significant ($P<0.05$). The study provides reference data for further screening and confirming *MYCN* gene as an anti-disease candidate gene.

【主要修改意见】

1. 注意可数名词的单复数形式，如"material"为可数名词，复数的形式为"materials"。
2. 合理分析汉语句子结构，选择恰当的中心词做英文句子的主语，如根据摘要的主题判断，原文汉语"本试验以大骨鸡为试验材料，采用半定量 RT-PCR 法对 MYCN 基因不同组织 mRNA 表达特性进行分析"中的核心信息是"MYCN 基因不同组织 mRNA 表达特性"而不是"采用半定量 RT-PCR 法"。

Unit 5　Sociology

3. 两个简单句间缺少连词，进行适当修改建立衔接关系，如"The results of PCR-SSCP showed that a variation C2178T was detected on conversion exon Ⅱ of *MYCN* gene, three genotypes were found：AA，BB，AB"中 that 从句包含两个并列分句，可加入 and 进行衔接。
4. 当并列使用两个以上的名词时，需用"and"连接，如"followed by lung, kidney, overy, brain"中的并列名词部分，应改为"followed by lung, kidney, overy and brain"。
5. 确保用词准确，如表达两个事实或数字之间的相关性时，应用"correlate" "correlation"来表述，而"associate" "association"是指某事物或某人在头脑里联想起其他事物或人。
6. 注意主谓一致，如"other immune parameters was not significant（$P<0.05$）"主语是复数形式，谓语动词应改为"were"。
7. 使用代词"it"指代研究结果，指代对象不明确，应将摘要结尾句中"it"改为"the study"。
8. 涉及实验研究意义的内容，建议使用一般现在时或一般将来时。如"It provided a reference data for us to further confirm *MYCN* gene act as an anti-disease candidate gene"中使用一般过去时，可改为一般现在时。
9. 注意译文应与汉语保持一致，忠实于原文。

Unit 6 Food Engineering

Text A

FDA 101: Product Recalls

Introduction

Everyday, a number of food companies are recalling their food items and interestingly the number is on rise. Reasons for these recalls are mostly presence of life threatening organisms like salmonella, undeclared allergens, and *Listeria monocytogenes* in the recalled items. Now, with incessant recalling of food products, questions that come into our mind are "Why so many food companies are recalling their food items?" "Why don't they check the safety of the food before releasing in market when they know unsafe food items can claim lives of many?"

Unsafe food can transmit disease from person to person as well as serve as a growth medium for bacteria that can cause food poisoning and take lives of many. To avoid this situation, food regulatory bodies have passed different regulatory measures, which ensure that food safety has been properly taken care of before releasing a food item in market.

In the U.S., The Food and Drug Administration publishes the *Food Code*, a model set of guidelines and procedures that assists food control jurisdictions by providing a scientifically sound technical and legal basis for regulating the retail and food service industries, including restaurants, grocery stores and institutional foodservice providers such as nursing homes.

There are 15 agencies sharing oversight responsibilities in the food safety system, although the two primary agencies are the U.S. Department of Agriculture (USDA) Food Safety and Inspection Service (FSIS), which is responsible for the safety of meat, poultry, and processed egg products, and the Food and Drug Administration (FDA), which is responsible for virtually all other foods.

A few regulations propagated by the FDA, can help the food producing companies eradicate the possibilities of recalling their marketed products.

FDA 101: Product Recalls

Once a product is in widespread use, unforeseen problems can sometimes lead to a recall. Contaminated spinach, for example, led to the recent recall of spinach products under multiple brand names. Contaminated peanut butter led to the recall of thousands of jars of two popular brands. In both cases, FDA responded immediately to minimize harm.

When an FDA-regulated product is either defective or potentially harmful, recalling that product—removing it from the market or correcting the problem—is the most effective means for protecting the public. In most cases, a recall results from an unintentional mistake by the company, rather than from an intentional disregard for the law.

Recalls are almost always voluntary. Sometimes a company discovers a problem and recalls a product on its own. Other times a company recalls a product after FDA raises concerns. Only in rare cases will FDA request a recall. But in every case, FDA's role is to oversee a company's strategy and assess the adequacy of the recall.

First Alert

FDA first hears about a problem product in several ways:

A company discovers a problem and contacts FDA.

FDA inspects a manufacturing facility and determines the potential for a recall.

FDA receives reports of health problems through various reporting systems.

The Centers for Disease Control and Prevention (CDC) contacts FDA.

When it comes to illnesses associated with food products, Dorothy J. Miller, Director of FDA's Office of Emergency Operations, says that FDA generally first hears of these kinds of problems from CDC.

"CDC hears about such problems from state health departments that have received and submitted illness reports," she says. "An ongoing outbreak means that we have an emergency, and when there's a public health crisis like this, you need to tell the public immediately."

Alerting the Public

FDA seeks publicity about a recall only when it believes the public needs to be alerted to a serious hazard. When a recalled product has been widely distributed, the news media is a very effective way to reach large numbers of people. FDA can hold press conferences, issue press releases, and post updates to its Web site regularly, to alert people.

"It's about being as transparent as possible," says Catherine McDermott, public affairs manager in the Division of Federal-State Relations in FDA's Office of Regulatory Affairs. "If we feel there is that much of a health risk, we will offer media updates every day to give new information, and all that we know gets posted to FDA's Web site."

Not all recalls are announced in the media. But all recalls go into FDA's weekly Enforcement Report. This document lists each recall according to classification (see "Recall Classifications" box), with the specific action taken by the recalling firm.

Effectiveness Checks

FDA evaluates whether all reasonable efforts have been made to remove or correct a product. A recall is considered complete after all of the company's corrective actions are reviewed by FDA and deemed appropriate. After a recall is completed, FDA makes sure that the product is destroyed or suitably reconditioned, and investigates why the product was defective in the first place.

Recall Classifications

These guidelines categorize all recalls into one of three classes, according to the level of hazard involved:

Class Ⅰ: Dangerous or defective products that predictably could cause serious health problems or death. Examples include: food found to contain botulinum toxin, food with undeclared allergens, a label mix-up on a lifesaving drug, or a defective artificial heart valve.

Class Ⅱ: Products that might cause a temporary health problem, or pose only a slight threat of a serious nature. Example: a drug that is under-strength but that is not used to treat life-threatening situations.

Class Ⅲ: Products that are unlikely to cause any adverse health reaction, but that violate FDA labeling or manufacturing laws. Examples include: a minor container defect and lack of English labeling in a retail food.

FDA-regulated Products Subject to Recall

human drugs

animal drugs

medical devices

radiation-emitting products

vaccines

blood and blood products

transplantable human tissue

animal feed

cosmetics

about 80 percent of the foods eaten in the United States

(967 words)

New Words

1. item ['aitəm] *n.* 物品
2. salmonella [,sɔlmə'nelə] *n.* 沙门氏菌
3. allergen ['ælədʒən] *n.* 过敏原
4. *Listeria monocytogenes* [lis'tiəriə ,məunəu'saitədʒi:ns] 单核细胞增多性李斯特氏菌
5. incessant [in'sesənt] *adj.* 持续不断的
6. release [ri'li:s] *vt.* 投放
7. claim [kleim] *vt.* 夺去生命
8. transmit [træns'mit] *vt.* 传播
9. code [kəud] *n.* 法典
10. jurisdiction [,dʒuəris'dikʃən] *n.* 司法管辖权
11. retail ['ri:teil] *n.* 零售
12. institutional [,insti'tju:ʃənəl] *adj.* 制度上的
13. agency ['eidʒensi] *n.* 机构
14. oversight ['əuvəsait] *n.* 监督
15. poultry ['pəultri] *n.* 家禽
16. virtually ['və:tʃuəli] *adv.* 几乎,实际上
17. propagate ['prɔpəgeit] *vt.* 宣传;繁殖
18. eradicate [i'rædikeit] *vt.* 杜绝
19. unforeseen [,ʌnfɔː'si:n] *adj.* 未预见到的,无法预料的

20. contaminate [kən'tæmineit] *vt.* 污染
21. spinach ['spinidʒ] *n.* 菠菜
22. defective [di'fektiv] *adj.* 有瑕疵的，有缺陷的
23. disregard [ˌdisri'ɡɑːd] *n.*，*vt.* 忽视
24. oversee [ˌəuvə'siː] *vt.* 监督
25. alert [ə'ləːt] *n.* 警报；*vt.* 向⋯发出警报
26. hazard ['hæzəd] *n.* 危害
27. update ['ʌpdeit] *n.* 更新的信息
28. enforcement [in'fɔːsmənt] *n.* 执行
29. review [ri'vjuː] *vt.* 审查
30. deem [diːm] *vt.* 视为，认为
31. recondition [ˌriːkən'diʃən] *vt.* 修复，修整
32. botulinum [ˌbɔtju'lainəm] *n.* 肉毒菌
33. toxin ['tɔksin] *n.* 毒素
34. mix-up ['miksʌp] *n.* 差错
35. pose [pəuz] *vt.* 造成，引起
36. under-strength ['ʌndə 'strenθ] *adj.* 药效不足的
37. violate ['vaiəleit] *vt.* 违反．
38. subject ['sʌbdʒikt] *adj.* 易受⋯影响的
39. vaccine [væk'sin] *n.* 疫苗
40. tissue ['tiʃjuː] *n.*（动、植物细胞的）组织
41. cosmetic [kɔz'metik] *n.* 化妆品

Useful Expressions

1. product recall　食品召回
2. regulatory body　监管机构
3. food safety　食品安全
4. manufacturing facility　生产设施
5. press conference　新闻发布会
6. press release　新闻公告
7. labeling law　标签法
8. medical device　医疗器械

Notes

1. 本文选自美国食品药品监督管理局（FDA）（www.fda.gov/consumer），文章全称：The Safety in the US。

2. 美国食品药品监督管理局（Food and Drug Administration，FDA），是隶属美国健康及人类服务部管辖的联邦政府机构，其主要职能是监管美国国内生产及进口的食品、膳食补充剂、药品、疫苗、生物医药制剂、血液制剂、医学设备、放射性设备、兽药和化妆品，同时也负责执行公共健康法案（Public Health Service Act）的第 361 号条款，包括公共卫生条件及州际旅行和运输的检查、对于诸多产品中可能存在的疾病的控制等。

3. 食品安全检验局（Food Safety and Inspection Service，FSIS），是美国农业部负责公众健康的机构，主要负责保证美国国内生产和进口的肉、禽类及蛋制品。

Exercises

Part Ⅰ Vocabulary and Structure

Section A Match each word with its Chinese equivalent.

1. sterilization   A. 过敏原
2. condense   B. 机制
3. allergen   C. 挥发性的
4. sanitation   D. 催化剂
5. catalytic   E. 原料
6. emulsify   F. 卫生
7. volatile   G. 灭菌
8. absorbent   H. 乳化
9. ingredient   I. 有吸收能力的
10. mechanism   J. 冷凝，浓缩

Section B Fill in the blanks with the words or expressions given below. Change the form where necessary.

| retail | contaminate | transmit | oversee | disregard |
| cosmetic | propagate | subject | incessant | correct |

1. As a result, our _____ interaction with food takes on immense power and can define who we are.
2. Finally, each year we will voluntarily report to Congress when we have invoked (行使) the privilege because there must be proper _____ of our actions.
3. Animal and vegetable pests _____ with extreme rapidity.
4. Environmental groups say the process can _____ drinking-water supplies, a charge the industry denies.
5. Some of the comments under the video displayed how a bias can make us _____ something without even looking at what it is.
6. So just one after another they made these insane departures (背离) from the _____ devices (手段) we'd put in the last time we had a big trouble and they really worked quite well.
7. The government agreed to peg down (约束) the _____ price of certain basic food stuffs.
8. The company has focused on stem cells, not only for treatment of chronic (慢性的) conditions, but for _____ .
9. Certain mosquitoes _____ malaria.
10. Clothing purchases over $200 are _____ to tax.

Part Ⅱ Translation

Section A Translate the following sentences into Chinese.

1. To avoid this situation, food regulatory bodies have passed different regulatory measures, which ensure that food safety has been properly taken care of before releasing a food item in market.
2. When an FDA-regulated product is either defective or potentially harmful, recalling that product is the most effective means for protecting the public.
3. When a recalled product has been widely distributed, the news media is a very effective way to reach large numbers of people.
4. This document lists each recall according to classification, with the specific action taken by the recalling firm.
5. After a recall is completed, FDA makes sure that the product is destroyed or suitably reconditioned, and investigates why the product was defective in

the first place.

Section B   Translate the following sentences into English.

1. 营养被消费者视为生活本身所必不可少的。
2. 这些生理效应是水溶性膳食纤维的典型效应。
3. 蛋白质能被分解产生不同大小和性质的中间物质。
4. 啤酒是一种麦芽经发酵获得的低酒精饮料。
5. 液体食物巴氏杀菌最通用的系统就是持续高温短时系统。

Text B

Food Security Strategies for Selected South Pacific Island Countries

The food security issue in South Pacific island countries is complex. The appropriate food security strategies and policy options can be *formulated*（制定；规划）properly only through a comprehensive study. Accordingly, the outcome of this project is intended to help the governments in South Pacific island countries to assess the extent of food insecurity, and devise appropriate food security strategies and formulate policy options. The general objectives of this project are to analyze food security conditions in selected South Pacific island countries and to formulate appropriate policy options for their food security strategies. The study was conducted using the latest "sustainable food security" *paradigm*（范式）, in which six criteria were analyzed for food security: food availability, access, utilization, stability, self reliance (autonomy) and sustainability. The countries that participated in the project are Fiji, Papua New Guinea, Tonga and Vanuatu. They were selected on the basis of similarities in traditional food staples (roots and tubers), dominant cultures (Melanesian and Polynesian), physical conditions and resource *endowments*（资源禀赋）, size, stage of economic development and geographical region. There is also a degree of diversity among the four countries for the purpose of contrast. All countries are ESCAP members that are infrequently invited to participate in CGPRT centre projects. Based on the framework, the main study subjects of the project are food security performance and its determinants at the national and household levels; food security risk-coping institutions; and feasibility of regional cooperation in food security.

The findings show that despite having limited *arable* (适于耕种的) land, a disadvantageous geographical location and space, and a small country size, and being prone to natural disasters, South Pacific island countries have managed to avoid acute food insecurity. All countries manage to *procure* (获得) sufficient food through domestic food production and importation. National food security is, however, in a potentially **precarious** condition in both the short run and long run. The major issue in the short run is temporary food insecurity due to vulnerability to various natural disasters, which are *endemic* (常见的) in South Pacific island countries. Through generations of experience, the people of these island countries have adapted well to their harsh living environment. They have developed various *indigenous* (本土的; 当地的) mitigation mechanisms, such as diversified and sequential farming systems, *egalitarian* (平等主义的) resource *tenurial* (土地保有的) systems, risk pooling social institutions (mutual-help organizations), indigenous food preservation techniques, wild food reservation areas and out-migration, effective enough to prevent acute temporary food insecurity induced by the endemic natural disasters. Perhaps, the most serious concern now is long-term sustainability of the national food security systems. The indigenous wisdom has been *eroding* (削弱) due to modernization processes and population pressure. Domestic food production capacity and productivity have shown declining trends and all countries have become increasingly dependent on food imports. Provincial and household food security is of more serious concern than national food security. Although the degree of segmentation varies by country, national food security follows a dualistic structure. Rural food security systems and urban food security systems are either separated or weakly related, chiefly due to deficiencies in marketing infrastructure. Food availability in rural areas primarily comes from local production, whereby access to food by household is determined by access to natural resources (arable land and *artisanal* (传统的) fishing grounds). As long as natural resources are abundant, rural food security systems remain strong and sustainable.

The most vulnerable provinces are those with high population pressure. The most vulnerable households are poor, with inadequate command over resources to produce subsistence foods and cash income. With the exception of Tonga, all countries studied are facing increased rural poverty that has become a serious threat to household food security in rural areas. One of the main causes in Papua New Guinea and Vanuatu is a high population growth rate, but Tonga and Fiji have managed to avoid this problem through emigration. Food availability in urban areas is heavily dependent on food importation. A household's access to food is determined by its purchasing power. The more serious problem in South Pacific island countries is nutritional insecurity. Both under-nutrition and over-nutrition are prevalent. Under-nutrition is caused by food insecurity or intra-household mal-distribution of foods among household members. Food insecurity is largely a poverty phenomenon, while intra-household mal-distribution of foods is a cultural phenomenon: husbands and older sons have first priority to access the foods available in the home. Women and children are the groups most vulnerable to under-nutrition, which is prevalent in Papua New Guinea, Vanuatu and Fiji where food insecurity is also prevalent. Over-nutrition is a syndrome of affluence that is prevalent among the middle to high income socio-economic groups due to over-eating of foods. Over-nutrition is highly prevalent in all South Pacific island countries and is arguably the most important issue of food security in the region. Strategy and policy recommendations for each case study country are elaborated in the respective country reports.

In general, the core issues that should be placed as the top priority of the national policy makers are: (i) Chronic food security faced by the poor household in both urban and rural areas of Papua New Guinea, Vanuatu and Fiji; (ii) Over-eating syndrome in all countries; (iii) Natural disaster induced temporary food security problems in all countries; (iv) Changes in traditional farming systems and their impacts on food security and resource sustainability; and (v) Social and *demographic* (人口统计学的) changes and their impacts on food security, priority and resource sustainability. South Pacific island countries have undertaken only limited trade of food commodities. The scope for regional cooperation includes: (i) Collaborative research and development on traditional crops that are common among the countries; (ii) Development of regional disaster preparedness and coping systems; and (iii) Development of regional agricultural research and development networks.

(933 words)

Comprehension of the Text

Choose the best answer to each of the following questions.

1. What aspect of study is not mentioned in the first paragraph?
   A. The problems to be solved through the study.
   B. The general objectives of the study.
   C. The paradigm used for conducting the study.
   D. The possible results of the study.
2. Which of the following statements is true?
   A. The outcome of this project is merely to help the governments in South Pacific island countries to assess the extent of food security.
   B. The general objectives of this project include analyzing food security conditions in South Pacific island countries.
   C. The study was conducted by using the latest "sustainable food security" paradigm with six criteria.
   D. All countries in the project are frequently invited to participate in CGPRT centre projects.
3. Which of the following is not the main study subject of the project?
   A. Food security performance.
   B. The determining factors at the national and community levels.
   C. Food security risk-settling institutions.
   D. Feasibility of territorial cooperation in food security.
4. Which of the following is not the factor in determining acute food insecurity of South Pacific island countries?
   A. Limited arable land.
   B. Geographical location and space.
   C. A small country size.
   D. Susceptibility to natural disasters.
5. What does the italicized and boldfaced word "precarious" in paragraph 2 probably mean?

A. Carious.   B. Precautious.
C. Insecure.   D. Previous.

6. Which of the following is the reason for temporary food insecurity?
   A. Susceptibility to various natural disasters.
   B. Insusceptibility to endemic natural disasters.
   C. People's harsh living environment.
   D. Various indigenous mitigation mechanisms.

7. What causes long-term sustainability of the national food security systems to be under threat?
   A. The indigenous wisdom has been prevailing.
   B. Domestic food production capacity and productivity have shown declining trends.
   C. All countries have become increasingly independent of food imports.
   D. Provincial and household food security has become more serious.

8. Which countries have a high population growth rate problem?
   A. Guinea and Vanuatu.
   B. Papua New Guinea and Vanuatu.
   C. Tonga and Fiji.
   D. Papua New Guinea, Vanuatu and Fiji.

9. What cultural phenomenon does intra-household mal-distribution of foods stand for?
   A. Food insecurity is largely a poverty phenomenon.
   B. Husbands and older sons have first priority to access the foods available in the home.
   C. Women and children are the groups most vulnerable to under-nutrition.
   D. Over-nutrition is a syndrome of affluence.

10. The scope for regional cooperation does not include _____.
    A. Cooperative research and development on traditional crops
    B. Development of regional disaster preparedness and coping systems
    C. Development of regional agricultural research and development networks
    D. Declining research and development on traditional crops

科技文体翻译技巧（六）

定义与描述的译法

在科技文献中，对事物外貌、结构、原理、过程等的描述一般始于定义，即指出事物的名称以及有别于其他事物的类属和特征。定义与描述这两个概念联系紧密。

一、定义

定义，是一种用简洁明确的语言对事物的本质特征做概括的说明方法。"定义"必须抓住被定义事物的基本属性和本质特征。

根据结构特点，科技文献中的定义被分为普遍定义、特指定义和扩展定义。基本表达方式和句型概况如下。

（一）普遍定义

普遍定义是一般陈述，被定义的事物首先用总类词描述，然后再用它的特有属性、功能、用途或起源描述。其结构有下述两个公式。

公式一：被下定义的事物（name）＋判断词＋种词（class）＋本质特性（characteristics）。

【例】Preservatives the natural or synthetic chemical composition, which is used in food, medicine, pigment, biological specimens, etc., to delay microbial growth or chemical changes caused by decay.

防腐剂是指天然或合成的化学成分，用于加入食品、药品、颜料、生物标本等，以延迟微生物生长或化学变化引起的腐败。

公式二：种词＋本质特征＋判断词＋名称。

【例】The process, through which water is lost from a plant primarily in the form of vapor, is known as transpiration.

水分从植物体以水蒸气状态散失的过程，称为蒸腾。

1. 功能定义

功能定义的可用表达公式：种词＋功能＋判断词＋名称；或名称＋判断词＋种词＋功能。

【例】Allele or (allelomorph) is a pair of alternative forms of a gene that can occupy the same locus on a particular chromosome and that control the same character.

等位基因一般指位于一对同源染色体的相同位置上控制着相对性状的一对基因。

2. 过程定义

过程定义用于说明生物体的生理过程、自然界的各种理化过程,以及生产操作或试验过程,下定义时可用公式:名称 ＋ 判断词 ＋ 种词 ＋ 识别性描述。

【例】 Metabolism in the body is the process whereby matter and energy exchange with organisms of the external environment as well as biological transformation process of material and energy.

生物体与外界环境之间的物质和能量交换以及生物体内物质和能量的转变过程称为新陈代谢。

3. 属性定义

属性定义用于说明物质的性质和成分构成。一般用下列公式表达:种词 ＋ 判断词 ＋ 性状。

【例】 Iron ore is the mineral aggregate containing iron or iron compounds with economic advantage. Iron is gradually selected from iron ore after crushing, grinding, magnetic separation, flotation, re-election and other procedures.

铁矿石是指含有铁单质或铁化合物、能够经济利用的矿物集合体。铁矿石经过破碎、磨碎、磁选、浮选、重选等程序逐渐选出铁。

(二) 特指定义

特指定义是指特定事物的定义,与表示一般事物的普遍定义相对应。从语义结构上讲,普遍定义结构中被下定义的通常是一个孤立的名词,而特指定义结构中被下定义的名词通常带有形容词、介词短语或其他定语。

【例】 普遍定义:Wilt is a disease which causes plants to drop and lose freshness.

枯萎病是一种造成植物枯萎和凋谢的疾病。

特指定义:Bacterial wilt in tomatoes is a bacterial disease caused by the organism *Pseudomonas solanacerum*.

番茄枯萎病是指由青枯菌引起的细菌性疾病。

(三) 扩展定义

在基本定义之后,为了概念更加清晰、消除误差、明确范围,有时要用到扩展定义,其目的是对基本定义进行补充说明。具体包括举例、对比、排除、罗列组成等手法。

1. 举例

通过实例更明确地限定事物的特征。

【例】Ammonia is a colorless, pungent, suffocating gas, NH_3, a compound of nitrogen and hydrogen, very soluble in water. Common examples are ammonia and liquid ammonia.

氨是一种无色,有强烈刺激性气味、令人窒息的气体,是氮和氢的化合物,极易溶于水。常用的实例有氨水和液体氨。

2. 对比

采用对比的方法可以使读者更好地把容易混淆的概念区分开来。

【例】Contagious disease is a disease that can spread by touch, while infectious disease is spread by air, water, etc.

接触性传染病是指通过接触传播的疾病,而传染性疾病是指由空气、水等传播的疾病。

3. 列出组成

下定义时列出组成部分,限定清晰的范围。

【例】Chemistry is the science to research the composition, structure, properties and the change rule of material at the atomic level, such as materials science, nanotechnology, biochemistry, etc.

化学是在原子层次上研究物质的组成、结构、性质及变化规律的科学,如材料科学、纳米科技、生物化学等。

4. 排除

用排除的方法,可以明确地把被下定义的事物与其他易混淆的事物区别开来。

【例】Insanity is the state of being mad. The condition called insanity does not include feeble mindedness, imbecility, or any other condition of mental deficiency as contrasted with mental derangement.

精神错乱是指疯癫的精神状态。被称为精神错乱的条件中不包括智力低下、低能或任何其他有别于精神错乱的智力缺陷。

5. 用途

扩展说明用途,可使定义更加清晰。

【例】Nitrogen fertilizer isa fertilizer containing crop nutrient element nitrogen, which can increase crop yield as well as play a significant role in improving quality of agricultural products.

氮肥是含有作物营养元素氮的化肥,它对提高作物产量,改善农产品的质量有重要作用。

（四）写定义的技巧

要用精练的语言完成一个概念清晰、范围准确且本质突出的定义，需要注意如下写定义的技巧。

1. 提炼"种词"

尽量缩小种词的概念范围。例如给 acid 下定义，种词要用 compound 而不用 chemical substance 或 one of a chemical substance。

2. 找准"特性词"

特性词是指能够揭示被定义事物与其他同类事物本质上差别的词，因此应避免概念不清楚或范围不明确。例如 acid 定义中的特性应当是 neutralizes a solution of sodium hydroxide 而不是其他。因此 acid 的完整定义是：An acid is a compound which neutralizes a solution of sodium hydroxide。

3. 正确使用谓语动词

因为普遍定义是一般的陈述，谓语动词用 is，不常用 be 的其他形式，有时用 can（may）be defined as。

二、描述

科技工作者要对很多事物进行描述，例如新育成作物品种的外貌、抗病虫害能力、产量等特点，特定区域土壤的肥力、酸碱度、降水量以及耕作手段等描述。实验的全部过程、用到的仪器设备、采用的实验方法、数据统计使用的软件等也是经常需要描述的内容。

科技描述与一般文学描述在文体上截然不同。文学文体允许在描述过程中加入作者的感情色彩，可以主观、夸张地突出事物某方面的特征。而科技文体则要求必须完全尊重事实，以客观严谨的态度，恰当准确地描述客观存在的事物或情况。下面以摘要中与实验相关部分及结论部分的描述为例，对科技文体描述予以说明：

实验方法及过程描述：In this work, a package was developed (thermosealed baskets) with table grapes wrapped with 2 distinct films (M and P) with different permeability (medium and high, respectively) without or with the addition of a mixture of eugenol, thymol and carvacrol.

本研究中，开发了一种包装（采用热封），对鲜食葡萄用两种通透性不同的膜（M 和 P）（分别为中等和高通透性）进行包装，并在包装袋中分别添加或不添加由丁香油酚、百里酚及香芹酚所组成的混合物。

实验结果描述：Table grapes stored on air (control) lost their quality attributes very rapidly, manifested by accelerated weight loss, color changes, softening and increase in soluble solids concentration and titratable acidity ratio (SSC/TA). Use of modified atmosphere packaging (MAP) alone retarded these changes, the effects being significantly greater when essential oils were added (especially for M film), although atmospheric composition was not affected by incorporating essential oils. In addition, microbial counts (fungi, yeasts and mesophilic aerobes) were decreased markedly and accompanied by a lower occurrence of berry decay. Although slight aroma was detected after opening the packages, absence of the typical flavor of these compounds was found by trained panelists after tasting the grapes.

对照组中（大气条件下）的鲜食葡萄很快失去其品质，伴随着加速的重量损失、颜色变化、组织变软及可溶性固形物（SSC）和可滴定酸（TA）的比值升高（SSC/TA）。气调包装（MAP）则能有效阻止这些变化，当添加上述精油后，尽管包装中大气的组成不会改变，但其效果仍然更佳（尤其对用中等通透性膜包装的）。此外，微生物的数量（真菌、酵母和嗜温性好氧菌）会显著减少，伴随较少的果品腐烂。尽管在打开包装后感受到些微的芳香物质，但专业品尝人员在品尝了这些葡萄后发现其缺失了其特征性风味物质。

结论描述：Results suggest that the overall quality (sensory and safety) of table grapes could be improved and the method considered an alternative to the use of synthetic fungicides.

这些结果说明了上述处理可以改进鲜食葡萄的总体品质（包括感官和安全性），并且可作为使用合成防霉剂的替代方法。

科技英语摘要写作（六）

摘要的结束句

摘要的结尾，通常有结束句，对全文做出结论或补充交代，指出其适用范围和有效性，或评价本研究的重要意义和实用价值，提出建议和措施。摘要结束句的时态上，一般性结论用一般现在时，特定条件下的结论用一般过去时，设想和建议用一般现在时。如下为常用句型句式。

1. The method discussed in this paper is simple, time saving and easy to understand. It may be used in wheat growing.

2. Consequently, there was no relationship between semi-dwarfing genes and amounts and pattern of water uptake in the dryland winter wheat cultivars investigated.

3. A thorough search of the World Oat Collection for more tolerant genotypes is recommended.

4. This paper argues that the changes in the above aspects will result in a technical revolution in the field of information security.

5. Experimental results have demonstrated that the malware signatures extracted show good ability to anti-obfuscation（抗干扰）and the detection based on theses signatures could recognize malware variants（恶意代码）effectively.

6. Finally, we use our analytical results to give managerial feedbacks for firms adopting and considering traceability.

7. Simulation results show that the proposed synchronization scheme performs well even at a low SNR（信噪比）and can effectively mitigate the effect of Rayleigh fading channel（衰落信道）.

8. It has proved that the model has advantages of high accuracy and fast convergence.

9. The hazard analysis is followed by exploring the cadmium limit in shellfish published in the current national standards, which can provide some information for the adjustment of the national standards and useful guidance for shellfish consumption.

10. Finally, this paper gives the future research directions, existing problems and challenges of DaaS（database as a service）in the security and privacy preserving.

摘要译写示例

示例一

【摘要中文原文】

摘要：本文建立一种利用双指纹图谱溯源进口霞多丽干白葡萄酒产地的方法。通过使用气相色谱-质谱（GC-MS）建立了进口霞多丽葡萄酒二氯甲烷提取物的特征指纹图谱,使用液相色谱-二极管阵列检测器-质谱(HPLC-DAD-MS)建立了进口霞多丽葡萄酒乙酸乙酯提取物的特征指纹图谱,此方法的精密度、重现性、稳定性好,符合指纹图谱评价技术的要求。利用 SPSS 19.0 软件对生成 20 种不同产地霞多丽干白葡萄酒的双指纹特征图谱进行聚类分析,分析结果能够准确地溯源进口霞多丽干白葡萄酒产地。本研究结果表明,将两种特征指纹图谱结合起来,通过多元数据分析可以实现快速、简单、有效溯源进口霞多丽干白葡萄酒的产地。

【摘要原版译文】

Abstract: A method of dual fingerprint analysis was established to trace the source region of imported chardonnay dry white grape wine. A GC-MS specific fingerprint of extractives with dichloromethane and a HPLC-DAD-MS specific fingerprint of extractives with ethyl acetate of imported chardonnay grape wine were established. The developed method was simple, accurate and highly sensitive, could meet the requirements of the fingerprint evaluation technology. Clustering analysis for 20 chardonnay dry white grape wine of different origins by SPSS 19.0 was used to trace their origins. The conclusion showed that the method combining the 2 kinds of specific fingerprint by using chemometric analysis could trace the origins of the imported chardonnay dry white grape wines fast, simply and effectively.

【摘要修改译文】

Abstract: A method of dual fingerprint analysis was created to trace source region of imported chardonnay dry white grape wine. A GC-MS specific fingerprint of extractives with dichloromethane and a HPLC-DAD-MS specific fingerprint of extractives with ethyl acetate of imported chardonnay grape wine were established. The developed method has advantages of high precision, good repeatability and stability, and could meet the requirements of the fingerprint evaluation technology. Cluster analysis for twenty kinds of chardonnay dry white grape wine of different origins by SPSS 19.0 was used to trace their origins. The results showed that the method combining the two kinds of specific fingerprint by using multivariate data analysis could trace the origins of the imported chardonnay dry white grape wines fast, simply and effectively.

【主要修改意见】

1. 第一句中"方法被建立"应理解为"方法被创建",因此,用"created"代替"established"。

2. 第三句中"此方法的精密度、重现性、稳定性好",原文翻译为"The developed method was simple, accurate and highly sensitive"选词不准确。应译为"The developed method has advantages of high precision, good repeatability and stability"。

3. 第三句应在"could"前加上连接词"and",平行结构中最后一项与前一项用"and"连接,使句子前后联系紧密,结构严谨。

4. 第四句,表示"聚类分析",用"cluster analysis"代替"clustering analysis"。

5. 最后一句,"conclusion"应换成"results",科技英语中表示实验结果用"results"而"conclusion"表示整体研究后得到的结论。

示例二

【摘要中文原文】

摘要:本文在料液比、浸提时间、浸提温度、pH 等单因素试验的基础上,采用正交试验法,优化大蒜抑菌物质的水提法工艺参数,优选出最优提取条件,利用抑菌试验探明不同条件下大蒜水提取物抑菌作用。试验结果表明,水法提取大蒜抑菌活性物质的最优提取条件为:料液比 1∶3,浸提温度 30 ℃,浸提时间 30 min,pH 6。在此条件下,大蒜水提取物对大肠杆菌的抑菌圈直径为 3.20 cm,说明其对大肠杆菌具有明显的抑菌作用。

【摘要原版译文】

Abstract: Based on the basis of single factor experiment like material liquid ratio、extraction time、extraction temperature、pH, using orthogonal test method to optimize the garlic bacteriostat qualitative water process parameters, select the optimal extracting conditions. Using bacteriostatic experiment to proven the bacteriostatic action of the water extract of garlic under different conditions. The test results shows that the water extraction of garlic antimicrobial active substances: the optimal condition for material liquid to 1∶3, leaching temperature 30 ℃, 30 min leaching time, pH 6. Under this condition, the water extraction of garlic on *E. coli* in the antibacterial circle diameter is 3.20 cm, instructions for the *E. coli* has obvious bacteriostatic action.

【摘要修改译文】

Abstract: Based on single factor experiments of material liquid ratio, extraction time, extraction temperature and pH, this study used orthogonal test method to optimize the extraction parameters of garlic bacteriostat. We selected the optimal extraction conditions and used bacteriostatic experiments to explore the bacteriostatic actions of the water extracts of garlic under different conditions. The test results showed that the optimal conditions for extracting active bacteriostatic substances with the above method were as follows: the ratio of material to liquid 1∶3, leaching temperature 30 ℃, leaching time 30 min, pH 6. Under these conditions, the inhibition zone diameters of water extracts to *E. coli* was 3.20 cm, which suggests the remarkable bacteriostatic action of the water extract on the *E. coli*.

【主要修改意见】

1. 第一句中"based on"与"the basis of"语义重复,表示"基于",可以用"Based on"或"On the basis of"表达。
2. 第一句"料液比、浸提时间、浸提温度、pH等单因素试验"应理解为"和料液比、浸提时间、浸提温度、pH等研究相关的单因素试验"所以用"of"不用"like"(like 科技英语中译为"诸如")。
3. 第一句中标点"、"不准确,英语中用","表示前后并列的多个事物。
4. 原译文前两句缺主语,造成句子结构不完整。当汉语表述中缺主语时,可考虑用被动语态或添加诸如"this paper""this study"这样的主语。
5. 表示一系列实验结果时可用"be as follows",这样可避免陈述中重复使用 be 动词。
6. 动词不定式"to"后接动词原型,"to proven"改为"to prove"。
7. 注意主谓一致,"the test results shows"中的"shows"应改为"show"。

Unit 7 Management

Text A

Structures of Project Portfolio Management

A major reason for the emerging project-oriented organization is the fact that firms nowadays run many projects simultaneously. Thus, there is an increasing need to coordinate and control complex project landscapes, in order to align projects to the strategic goals, to pick the winners, to avoid an accumulation of risk, to manage synergies between projects, to adapt to changes, and to provide project teams with sufficient resources and to avoid work overload. This is usually done in project portfolio management. The required strategic and operational transparency is established, if processes and structures for project portfolio management are well organized, if planning and control instruments are established professionally, and if both functions are supported by information systems with a high utility and usability. Thus, project portfolio management contains three components: (a) organizing of structures and processes, (b) planning and controlling, and (c) Information and Communication Technology (ICT) systems to support decision-making.

Organization of Structures and Processes

Decisions about project portfolios are often made by specific project portfolio boards. An organization may have different kinds of portfolios, and establish a portfolio board for each kind, or for different organizational parts, and it may also establish higher order boards governing the decisions, which portfolio boards make. Often the work of such a project portfolio board is supported by a project portfolio management office, which performs coordinating, planning and controlling, and supportive functions and increases project performance. In addition to this, project-oriented organizations may establish expert units and project leader units. People, who often work in projects, thus get a home-base in the permanent organization adapted to their specific needs. They interact more intensively and share knowledge between projects, and they get a project-oriented manager, who takes care of their careers in projects, and supports them if conflicts arise that are typical in project work, or that arise because people are working in several projects simultaneously. The permanent organization has a better overview of its project specialists, and can assign them more easily to projects.

Various studies support the notion that the formalization of project portfolio processes significantly influences portfolio performance. These formal processes introduce structure, sequence, and clarity to all projects. Establishment of clear rules and guiding principles at the decision points lead to data integrity and facilitate the comparison of divergent projects ensuring that processes are comprehensive and responsibilities are well defined. The formalization of stages of project portfolio process into project portfolio structuring, resource allocation management, portfolio steering, and exploitation and competence securing has repeatedly shown to increase project portfolio performance.

Planning and Controlling

A project-oriented organization is a future-oriented organization, because projects are intended to improve our future. Such a future-oriented organization requires that the organization develops a well-founded viable strategy, which is broken down to the project portfolio level, because a company's strategy is realized by the entirety of its projects. This means that operational criteria are developed, which allow to align the project portfolio with the organizational strategy. Given correct information about the projects and the resource base, this should enable the portfolio boards to prioritize the best projects and to terminate those that make no sense. Strategic clarity is thus the first condition for planning and controlling project portfolios. However, in increasingly turbulent environments it is necessary, but not sufficient to rely on deliberate strategies.

A second condition for project portfolio success is operational clarity about the projects, their expected benefits, risks, and resource requirements; the resources and their quality and availability. Jonas et al. present a new construct called "management quality", which comprises three facets: cooperation quality, information quality, and allocation quality. Cooperation quality measures the quality of cross-project cooperation between different project managers and project teams. The conceptualization of information quality is influenced by DeLone and McLean and comprises relevance, understandability, accuracy, conciseness, completeness, currency, timeliness, and usability of the information, which the decision makers on the portfolio board and the project portfolio manager can access. A high information quality helps to allocate resources better according to value creation, risk and strategic goals, and it speeds up decision-making processes. Allocation quality measures the effectiveness, speed, stability, and conflict handling quality of human resource allocation decisions. These three facets capture portfolio management on a meta-level. Kopmann et al. measure business case control as a second-order construct, which comprises the requirement of business cases for assessing of all projects, monitoring and adapting these business cases over the project life cycle, and tracking the success of projects in the portfolio for a considerable time after their completion. The investigation shows a significant positive effect on project portfolio success. This influence increases if project managers, line managers, and project portfolio managers are incentivized for project portfolio success. The positive influence of business case control also increases with portfolio size and complexity, as well as environmental turbulence.

Risk management is also a major theme in planning and controlling projects and project portfolios. The study from Teller and Kock identifies two components of risk management at the portfolio level: (1) Creating transparency about risks and (2) Establishing capacity to cope with risks. Transparency is fostered by risk identification activities, by a formalized risk management process, and by a culture which fosters a frank and open communication about risks. Risk-coping capacity is increased by risk prevention, risk monitoring, and integration of risk management in project portfolio management. Transparency about risks and capacity to cope with risk show a significant positive influence on project portfolio success. Teller et al. show that the integration of the information about risks expected in single projects into a project portfolio risk assessment is crucial for project portfolio success, and that the positive influence of this integration increases with increasing environmental turbulence.

ICT Systems to Support Decision-making

Information and communication systems are nowadays ubiquitous in project work and project management. The focus here is on ICT systems to support the management of single projects and project portfolios, in particular for systems to support planning, controlling and coordinating functions, and decision-making. Computer-aided support for making decisions in projects has a long tradition. Tools like CPM or PERT, developed in the 1950ies helped to plan and schedule complex projects more efficiently and contributed much to the diffusion of project management tools and practices. One may conclude that organizations striving for project-orientation should first do their homework and formalize their processes in order to get better information into their systems, and to avoid garbage-in-garbage-out decision support systems. However, on the other hand the implementation of Project Management Information Systems (PMIS) and Project Portfolio Management (PPM) systems are useful in defining the PM (Project Management) processes better and to lay more stress on information quality and utility and usability of decision support systems. In addition, there should be a fit between the processes, the decision-making culture and the decision support systems.

Overall, the three structural components (1) organizing of specific integrative permanent structures, of project portfolio processes and of roles, (2) implementation of an integrated project planning for single projects and project portfolios and its alignment with strategic planning, and (3) ICT systems to support and automate these tasks, have been shown to increase the maturity level of project management, and the performance of single projects and project portfolios. If the strategy is to create more value through more innovative products, services, processes, and infrastructures, the structural components can foster such a goal. If an organization classifies their projects according to their innovativeness, and if it reserves specific budgets within which highly innovative projects only compete with other highly innovative projects for scarce resources, such a resource shield may protect exploratory projects, which can generate real options for future follow-up projects.

The contributions of key people like promoters, technological gatekeepers, brokers and stewards, have often been shown to overcome barriers against innovation and develop and market innovations successfully. The implementation of the above described structures, which can be considered a management innovation, requires an active engagement of such people. In the organizational development process to master this change, the key persons may act as mentors, coaches, peers, networkers, and sponsors, which bring required people together, create confidence and trust in the value of this management innovation, and legitimize required investments of financial and human resources.

(1,402 words)

New Words

1. simultaneously [ˌsiməl'teiniəsli] *adv.* 同时
2. coordinate [kəu'ɔːdineit] *vt.* 使协调；使相配合
3. align [ə'lain] *vt.* 调整，使一致
4. strategic [strə'tiːdʒik] *adj.* 战略上的
5. synergy ['sinədʒi] *n.* 协同作用
6. transparency [trænsˈpærənsi] *n.* 透明度
7. utility [juːˈtiləti] *n.* 实用，效用
8. intensively [in'tensivli] *adv.* 广泛地
9. assign [ə'sain] *vt.* 分配
10. formalization [ˌfɔːməlaiˈzeiʃn] *n.* 正规化，公式化
11. integrity [in'tegrəti] *n.* 完整性
12. facilitate [fə'siliteit] *vt.* 促使；使便利
13. divergent [daiˈvəːdʒənt] *adj.* 不同的；有分歧的
14. allocation [ˌæləˈkeiʃn] *n.* 分配，配置
15. steering ['stiəriŋ] *n.* 筹划指导
16. exploitation [ˌeksplɔiˈteiʃn] *n.* 开发；利用
17. viable ['vaiəbl] *adj.* 可行的
18. criteria [kraiˈtiəriə] *n.* 标准
19. prioritize [praiˈɔrətaiz] *vt.* 按优先顺序；优先处理
20. terminate ['təːmineit] *vt.* （使）停止，终止
21. turbulent ['təːbjələnt] *adj.* 动荡的；混乱的
22. deliberate [diˈlibərət] *adj.* 深思熟虑的
23. construct [kənˈstrʌkt] *n.* 结构体
24. facet ['fæsit] *n.* 方面，部分
25. conceptualization [kənˈseptjuəlaiˈzeiʃən] *n.* 构思；观念的形成

26. conciseness [kən'saisnis] n. 简洁
27. currency ['kʌrənsi] n. 流通
28. timeliness ['taimlinəs] n. 及时性
29. incentivize [in'sentivaiz] vt. 刺激；激励
30. foster ['fɔstə(r)] vt. 促进
31. frank [fræŋk] adj. 坦诚的
32. ubiquitous [juː'bikwitəs] adj. 十分普遍的
33. diffusion [di'fjuːʒn] n. 传播；散播
34. implementation [ˌimplimen'teiʃn] n. 实施
35. automate ['ɔːtəmeit] vt. 使自动化
36. alignment [ə'lainmənt] n. 一致性
37. broker ['brəukə(r)] n. 经纪人
38. steward ['stjuːəd] n. 统筹人
39. mentor ['mentɔː(r)] n. 顾问
40. legitimize [li'dʒitəmaiz] vt. 使合法化；正式批准

Useful Expressions

1. project portfolio   项目组合
2. project landscape   项目景观
3. work overload   超负荷工作
4. expert unit   专家团队
5. resource allocation   资源配置
6. competence securing   能力的获得
7. on a meta-level   在元（模型）层
8. environmental turbulence   环境动荡
9. risk identification   风险识别
10. risk assessment   风险评估
11. lay stress on   重视；强调
12. resource shield   资源保护

Notes

1. 本文选自 *International Journal of Project Management* 期刊 2017 年

的论文 *The project-oriented organization and its contribution to innovation*，作者为 Hans Georg Gemünden，Patrick Lehner，Alexander Kock。

2. 信息和通信技术（Information and Communication Technology，ICT），是电信服务、信息服务、IT 服务及应用的有机结合，这种表述更能全面准确地反映支撑信息社会发展的通信方式，同时也反映了电信在信息时代自身职能和使命的演进。

3. 关键路径法（Critical Path Method，CPM），是一种网络图方法，由雷明顿兰德公司（Remington Rand）的克里（JE Kelly）和杜邦公司的沃尔克（MR Walker）在 1957 年提出，用于对化工工厂的维护项目进行日程安排，适用于有很多作业，而且必须按时完成的项目。关键路径法是一个动态系统，它会随着项目的进展不断更新。该方法采用单一时间估计法，其中时间被视为一定的或确定的。

4. 网络分析法（Program Evaluation and Review Technique，PERT），即计划评估和审查技术。网络分析法是利用网络分析制订计划并对计划予以评价的技术。它能协调整个计划的各道工序，合理安排人力、物力、时间、资金，加速计划的完成。在现代计划的编制和分析手段上，被广泛使用，是现代化管理的重要手段和方法。

Exercises

Part I Vocabulary and Structure

Section A Match each word with its Chinese equivalent.

1. deliberate   A. 促进
2. facet   B. 倡导者，支持者
3. mentor   C. 方面，层面
4. foster   D. 战略上的
5. synergy   E. 深思熟虑的
6. currency   F. 结构体
7. strategic   G. 顾问
8. construct   H. 协同作用
9. promoter   I. 使自动化
10. automate   J. 流通

Section B Fill in the blanks with the words or expressions given below. Change the form where necessary.

| diffusion | legitimize | turbulent | align | utility |
| incentivize | conciseness | divergent | landscape | prioritize |

1. Politeness, clearness and _____ are the essentials of good business writing.
2. In areas where the wind dies down, backup electricity from a _____ company or from an energy storage system becomes necessary.
3. As businesses look to _____ activity within their internal or external networks, they may include carrots that encourage a bit of friendly competition.
4. It can be hard to identify which parcels are carrying crucial items, but USPS (美国邮政署) and UPS (美国联合包裹运送服务公司) try their best to _____ sensitive material.
5. They had been together for five or six _____ years of break-ups and reconciliations.
6. The enterprise architect role should _____ business and IT requirements in service-oriented architecture.
7. They will accept no agreement that _____ the ethnic division of the country.
8. There are widely _____ views on whether Chinese stocks are currently cheap or expensive.
9. The _____ of knowledge accelerated dramatically in recent decades.
10. This pattern of woods and fields is typical of the English _____ .

Part Ⅱ Translation

Section A Translate the following sentences into Chinese.

1. Often the work of such a project portfolio board is supported by a project portfolio management office, which performs coordinating, planning and controlling, and supportive functions and increases project performance.
2. Establishment of clear rules and guiding principles at the decision points lead

to data integrity and facilitate the comparison of divergent projects ensuring that processes are comprehensive and responsibilities are well defined.

3. Such a future-oriented organization requires that the organization develops a well-founded viable strategy, which is broken down to the project portfolio level, because a company's strategy is realized by the entirety of its projects.

4. One may conclude that organizations striving for project-orientation should first do their homework and formalize their processes in order to get better information into their systems, and to avoid garbage-in-garbage-out decision support systems.

5. The contributions of key people like promoters, technological gatekeepers, brokers and stewards, have often been shown to overcome barriers against innovation and develop and market innovations successfully.

Section B　Translate the following sentences into English.

1. 管理的基本原理在人类文明之初就已经存在了。
2. 投资者对公司的潜力必须做出有依据的猜测。
3. 我们将有能力为客户提供更优质的产品，并且为股东带来更丰厚的回报。
4. 根据此项分析，最富裕成年人的收入增幅是中等收入者的两倍。
5. 员工管理的重要任务之一是培养有利于吸引、发展和留住人才的创新能力。

Text B

Five Principles of Great Management

According to Steve Jobs, "Simple can be harder than complex: You have to work hard to get your thinking clean to make it simple." By understanding and learning to apply these universal principles, you are more likely to excel as a manager in any organization.

Principle No. 1: The Functions of Management

While managers often view their work as task or *supervisory*（管理，监控）in orientation, this view is an *illusion*（错误的观念）. At the most fundamental level, management is a discipline that consists of a set of five general functions: planning, organizing, staffing, leading and controlling. These five functions are part of a body of practices and theories on how to be a successful manager.

Understanding the functions will help managers focus efforts on activities that gain results.

 a. Planning: When you think of planning in a management role, think about it as the process of choosing appropriate goals and actions to pursue and then determining what strategies to use, what actions to take, and deciding what resources are needed to achieve the goals.

 b. Organizing: This process of establishing worker relationships allows workers to work together to achieve their organizational goals.

 c. Leading: This function involves *articulating* （明确有力地表达） a vision, energizing employees, inspiring and motivating people using vision, influence, persuasion, and effective communication skills.

 d. Staffing: *Recruiting* （招聘） and selecting employees for positions within the company (within teams and departments).

 e. Controlling: Evaluate how well you are achieving your goals, improving performance, taking actions. Put processes in place to help you establish standards, so you can measure, compare, and make decisions.

Principle No. 2: The Types and Roles of Managers Within the Organization

Organizational structure is important in driving the business forward and every organization has a structure. No matter the organizationally specific title, organizations contain front-line, middle, and top managers. Above the top management team are a CEO and a board of director levels. To see this structure even more clearly, *visualize* （想象） a pyramid model. The more you move toward the top of the pyramid, the fewer managers you have. All of these management roles have specific tasks and duties. According to Jones and George, "A managerial role is the set of specific tasks that a manager is expected to perform because of the position he or she holds in an organization." These skills can be gained with a degree in organizational management.

All great managers play important roles in this model. One important thing to remember is from Henry Mintzberg, a management scholar who researched and reduced thousands of tasks performed by managers to 10 roles.

His model points out that there are three main types of roles all managers play; they are decisional, interpersonal, and informational. In the decisional role, managers can perform in an entrepreneurial manner, as a disturbance handler, resource allocator or negotiator. In an interpersonal role, managers may be figureheads, leaders, and *liaisons* (联络人). In the informational role, they monitor, are *disseminators* (传播者) or spokespersons, and they share information.

Principle No. 3: Effective Management of Organizational Resources

An essential component of operationalizing the organization's strategic plan is allocating resources where they will make the most impact. In fact, Dr. Ray Powers, associate dean in the Forbes School of Business & Technology (FSB), argues that it is the most important thing to do.

"I define resources as people, time, money, and assets—and of course the basic definition of a project is to have a goal and a start and end date—for pretty much any activity we do," he explains.

Managers participate in operational planning and budget planning processes and, in doing so, actively determine what should be done, in what order it is to be done, and determine what resources are appropriate to be successful in achieving the plan. Keep in mind that this is not a personality contest. The strategic plan and its specific objectives determine what is important and what may not be as important.

Principle No. 4: Understanding and Applying the Four Dimensions of Emotional Intelligence (EQ) in Maximizing Human Potential

Effective managers understand the context and culture in leadership situations. What helps these managers succeed? It is simple; they understand EQ (the competencies in each dimension of emotional intelligence).

Those four dimensions are: a high self-awareness, social awareness, self-management, and good social skills. All of these competencies are important, and they lead to great connections with people. They lead to stronger and more effective managerial performance. EQ is a very important component for excelling as a supervisor.

The job of the manager is to find a way to turn a team member's skill and talent into a higher level of performance. This idea doesn't suggest **manipulation** at all. Instead, it is about maximizing human potential, one team member at a time. It is as much art as it is science.

Dr. Diane Hamilton, program chair in the Forbes School of Business & Technology, recently described a candidate seeking a position on the faculty senate with a high EQ. Dr. Hamilton, a highly skilled professional who possesses knowledge and skill in the area of Myers Briggs Type Indicator, recognizes the importance of EQ.

"He demonstrates emotional intelligence and exemplifies the high *caliber* (才干) of candidate I would like to represent the FSB," she said about the candidate.

Principle No. 5: Know the Business

A common *axiom* (格言，定理) in management is that a qualified manager can manage any business. This point is only partially true. It is true that most managers are generalists rather than specialists; however, many very successful managers began their careers in specialist roles. What most successful managers bring to their work in leading crews, departments, divisions, and companies is both a solid knowledge of the business (they are very experienced) and a solid knowledge of the principles of great management. Manager *aspirants* (有抱负的人) must first learn the characteristics of the business by doing, working *in the trenches* (在第一线), and discovering how the various pieces of the organization work together to become a universal whole because very good managers discover what is universal in the business and capitalize on it to advance the business and improve performance.

Conclusion

Remember, as a manager, for greater job satisfaction and career success you should align to your organization's vision, mission, strategies, leadership, systems, structure, and cultures. In all you do, treat people fairly and honestly and do your best to follow and embrace your organization's ethics and core values as well as your own. **Talk the walk and walk the talk**, and remember, people are watching and seeing how you walk it. Give your very best to your teams, organizations, and customers. Be an effective manager to get the performance results for your organization and build trust and positive relationship with your people.

(1,123 words)

Unit 7　Management

Comprehension of the Text

Choose the best answer to each of the following questions.

1. Which of the following statements best summaries the text?
   A. For managers, simple things are harder than complex things.
   B. By understanding the principles of management, people can definitely become a successful manager.
   C. Successful managers conform to the company's vision, mission, strategies, leadership, systems, structure, and cultures.
   D. People are watching the managers; therefore, they should be very careful in their jobs.
2. We can infer from the passage that _____.
   A. managers' work is supervisory that need to be based on the functions of great management
   B. management consists of fundamental tasks that help managers focus their efforts on productive activities
   C. the five management principles are the whole body of the practices and theories on how to become a great manager
   D. by taking their work as supervisory in orientation, the managers intend to become illusionists
3. What function involves stimulating employees?
   A. Planning.　　　　　　B. Organizing.
   C. Staffing.　　　　　　D. Leading.
4. By citing the words of Jones and George, the author intends to show that _____.
   A. a manager's position in an organization decides his responsibilities
   B. a managerial role is the fundamental thing in leading a company
   C. the higher you climb the corporate ladder, the less managers you'll have
   D. the set of specific tasks decides the organization's future direction
5. In the informational role mentioned by Henry Mintzberg, the managers _____.
   A. act as information providers
   B. act as the speaker of all managing talents

C. act as the resource allocators

D. act as the leader of the project

6. According to Dr. Ray Powers, _____ .

    A. resource refers to all the materials in our environment which help us to satisfy our needs

    B. the definition of management refers to the optimal way to accomplish tasks and achieve goals

    C. to operationalize the strategic plan, managers just need a goal

    D. the most important thing in managing is to allocate resources reasonably

7. Which of the following is NOT true according to the passage?

    A. Managers determine what should be done.

    B. Participating in planning is not a personality contest for the managers.

    C. EQ, is a measure of a person's reasoning ability.

    D. Effective managers understand EQ of candidates.

8. What does the word "manipulation" (Line 2, Para. 11) mean?

    A. The making of articles on a large-scale using machinery.

    B. A book, document, or piece of music written by hand rather than typed or printed.

    C. Joy or satisfaction resulting from a success or victory.

    D. The action of controlling someone or something by artful, unfair, or insidious means.

9. The citation of "a qualified manager can manage any business" in "Principle No. 5" is to illustrate that _____ .

    A. great managers could get a solid knowledge of the management principles

    B. people tend to be biased when it comes to the idea of what qualifies someone to be a manager

    C. a qualified manager is capable of running any type of business

    D. all successful managers begin their careers in specialist roles

10. What does "talk the walk and walk the talk" mean in the last paragraph?

    A. Believing in your organization's values will help you set up your own ethical rules.

    B. Managers need to match their words with their actions consistently.

    C. Since people are watching how the managers walk, they need to be attentive in talking.

D. Managers need to build trust with his people by sharing some healthy ideas.

科技文体翻译技巧(七)

实验与比较的译法

一、实验

科技文体写作中通常包含介绍实验的部分,如实验目的、内容、过程、方法、结果、结论等,这些部分的英文表达在科技文体翻译中尤其重要。有一系列句型句式可以表达这些概念,举例如下:

(一) 实验目的

【例】1. Scientific research is made for the purpose of....
科学研究的目的是……。

2. The aim of (in doing) our scientific research is to realize....
我们(所做)的科学研究目标是实现……。

3. Experiments on ... were carried out to determine some questions about ....
进行有关……实验是为了确定一些有关……的问题。

(二) 实验结果

【例】1. The present series of experiments on ... showed a variety of changes in ....
目前的一系列关于……的实验表明……的各种变化。

2. The treatment conducted on ... resulted in....
关于……的处理导致……。

3. These results obtained are incompatible with the findings reported by....
得到的这些结果与……报道的结论不一致。

(三) 实验结论

【例】1. As a result of our experiments we conclude that....
作为实验结论,我们推断出……。

2. From these experiments we came to realize that....
   从这些实验当中我们认识到……。
3. The problem on ... demands further investigation in this area.
   有关……的问题要求在这个领域进一步调查研究。

二、比较

科技文体翻译的另一个重要部分是如何正确地用英语表达实验中相比较的两个事物。

（一）对比

对比是表示两个事物不同，常用 while 或 whereas 连接。

【例】X 在高温下是更有效的，而 Y 在低温下更有效。

这句话有如下表达法：

1. At high temperature X is more effective, while at low temperature Y is more effective.
2. Whereas at high temperature X is more effective, at low temperature Y is more effective.

（二）相似

相似表示两个事物的相似程度。

【例】This substance is exactly the same as/similar to the other one.
这种物质与另一种完全相同/相类似。

（三）区别

区别表示两个事物有何区别和差异程度。

【例】The plant can be distinguished from the other one in several respects.
这种植物与另一种有几方面的区别。

（四）最高和最低

这是表示两个相比较事物的最高程度或最低程度。

【例】The maximum/minimum temperature in this experiment is about 30℃.
这个实验的最高/最低温度约为 30℃。

（五）过多和过少

这是表示两个相比较事物的程度过分或不足。

【例】The excessive amount of air increases the temperature of fuel.
过多的空气会提高燃料的温度。

科技英语摘要写作（七）

摘要缩写（Ⅰ）

摘要是报告、论文内容不加注释和评论的简短陈述。英文摘要的内容要求与中文摘要一样，基本要素包括目的（objective）、方法（methods）、结果（results）和结论（conclusion）4 部分。具体地讲就是研究工作的主要对象和范围、采用的手段和方法、得出的结果和结论，有时也包括具有情报价值的其他重要信息。

缩写英文摘要应注意简洁明了，力争用最精练的语言和最短的篇幅提供最主要的信息。缩写英文摘要时应遵循的基本原则：①对原英文摘要资料进行精心筛选，不属于上述 4 个基本要素的内容不必写入。②对属于基本要素的内容，也应适当取舍，做到简明扼要，不包罗万象。比如"目的"，在多数标题中已初步阐明，若无更深层的目的，缩写英文摘要时完全不必重复叙述。再如"方法"，在缩写英文摘要时写出方法名称，即可不必描述其具体操作步骤。③缩写英文摘要并不是在原英文摘要中随意删去一些内容，把长文变短文，那样很容易造成摘要重心转移，甚至偏离主题。缩写英文摘要时也应严格地遵循英文语法修辞规则，符合英文专业术语规范，并兼顾英文的表达习惯。

缩写英文摘要主要有两种方法：提取要素法和综合概括法。这里介绍第一种方法。

提取要素法

把原英文摘要中目的、方法、结果和结论 4 个要素加工提炼，浓缩为能突出表现原摘要核心内容的段落。举例如下：

Inquiry into the Essence of Ideological and Political Education

【原摘要】

The essence of ideological and political education is a basic question in terms of theory and in practice. To realize the question, we can use Marxist methodology to implement the process of understanding step by step and construct the logic system of the theory of the essence of ideological and political education. One way to understand the theory of the essence of ideological and political education is to understand its meaning, philosophy and content. It has especially important and practical meaning for the use of the theory of the essence of ideological and political education in the new condition to cognize and deal properly with the relationship between its nature of class and science, centralization and variation, constancy and fluctuation.

【缩写后摘要】

The aim of this study is to realize the essence of ideological and political education. The scope of the research covers the theory on philosophy and methodology. The result reveals the realistic significance of ideological and political education.

摘要译写示例

示例一

【摘要中文原文】

摘要：本文论述了监控预警技术中的网络搭建、数据实时采集与无线传输、手机短信预警信息发布与远程控制的技术方法。该方法实现了对不同传感器协议的兼容及组网技术，通过远程服务器对数据进行统一管理，用户通过计算机客户端、web客户端和手机客户端可实现远程实时监测，若数据异常时，可通过客户端、手机短信等预警方式实时预警，并可控制远程设备，实现灌溉、通风等。该技术已在吉林省进行示范应用，有效地解决了设施农业集中化管理，提高了设施农业园区管理水平和应对异常的能力。

【摘要原版译文】

Abstract: This paper discussed the technology and methods that include construction of monitoring network for the greenhouse and other growing environment, real-time data acquisition and wireless transmission of data, early warning information release and remote control of the equipment. By different sensor protocol compatibility and networking technology, used the remote server data unified management, users logged in through the computer client, the web client, and the mobile phone client, realize remote real-time monitor environment of the facilities, If the data is abnormal, may achieve the real-time warning through the client, mobile phone short messages warning, control remote devices realize irrigation and ventilation. The technology has been applied to Jilin province demonstration, and effectively solve the centralized management, improve the level of facility agriculture park management and the ability to deal with abnormal.

【摘要修改译文】

Abstract: This paper discussed the techniques on network construction of monitoring and warning, real-time data acquisition and wireless transmission of data, early warning information release of mobile phone short messages and remote control . By different sensor protocol compatibility and networking technology, using the remote server data unified management, users can log in through the computer client, the web client, and the mobile phone client to realize remote real-time monitor. If the data are abnormal, the users may achieve the real-time warning through the client, mobile phone short messages warning and realize irrigation and ventilation by controlling remote devices. The technology has been applied in Jilin province for demonstration, and it can effectively achieve centralized management and improve the level of facility agriculture park management and the ability to deal with abnormal situations.

【主要修改意见】

1. 科技文体语言力求简练和准确。使用"technology and methods"来表达"技术方法"略显中式英语复杂化，可简译为"techniques"。
2. 句型表达也应简化。定语从句"that include construction of..."可以用"on network construction of"短语的形式表达。
3. 中英译文应对应。中文摘要中未出现"the greenhouse and other growing environment"和"environment of the facilities"的表达，此短语可省略。
4. 科技用词精准化。"手机短信"不能简单译为"equipment"，将其改为"mobile phone short messages"。
5. 注意非谓语动词主被动形式的准确使用。当主语为"users"时，非谓语动词"used"应译为"using"，表示"用户"主动发出的动作。
6. 避免连动形式。"users logged in... realize..."属于错误的连动表达，应改译为"... to realize..."，以不定式的形式表达原文。
7. 保持主谓一致。"data is"应译为"data are"。
8. 正确给出条件从句中的主语。句子"If the data..., may achieve the real-time..."应添加主语，改译为"If the data..., we may achieve the real-

time."。
9. 规范介词的使用。"applied to Jilin province demonstration"中的介词表达不规范，应译为"applied in Jilin province for demonstration"。

示例二

【摘要中文原文】

摘要：针对玉米剥皮机测试平台的角度控制系统，提出了基于电动推杆的角度控制系统。控制系统采用 PID 控制的方式，通过伺服驱动器精确控制电动推杆的移动距离和定位。系统采用双伺服驱动闭环控制、增量式 PID 位置闭环控制以及临界工作点保护控制方式相结合，并通过上位机软件实现对两个伺服驱动器的同步控制，经过测试运行精确可靠，实现了玉米剥皮机测试平台的角度同步调整和精确控制。

【摘要原版译文】

Abstract：This paper mainly focuses on the angle control system of corn husker test platform, and presents an angle control system based on linear actuator. PID control method is used in the electric control system, the movement distance and location of the linear actuator is also accurately controlled by servo driver. Adopting double servo closed-loop control, and combines with incremental PID position closed-loop control and critical point protection control method, the synchronization control of two servo driver is carried out by upper computer software. The test result show that system runs accurately and reliably, and implement the angle synchronous adjustment and precise control of the corn husker test platform.

【摘要修改译文】

Abstract：This study mainly focuses on the angle control system of corn husker test platform, and presents an angle control system based on linear actuator. PID control method is used in the electric control system. The movement distance and location of the linear actuator are also accurately controlled by servo driver. By combining double servo closed-loop control, incremental PID position closed-loop control and critical point protection control methods, the system realizes the synchronization control of two servo drivers with upper computer software. The test results show that the system runs accurately and reliably, and implements the angle synchronous adjustment and precise control of the corn husker test platform.

【主要修改意见】

1. 规范词语表达。研究属于实验类别，使用"study"较"paper"更符合科技文体表达。
2. 句子结构清晰化。原版译文"PID control method is used…, the movement distance and location of the linear actuator is also accurately controlled…"中出现了两个简单句相连的错误结构。一般可以使用句号断句或"and"连接。
3. 简化翻译，避免连动。原版译文"Adopting double servo closed-loop control, and combines with incremental PID…"中出现连动的语法错误，可以简译为"By combining double servo closed-loop control, incremental PID…"。
4. 注意单复数的使用。短语"protection control method"中的"方式"应为复数含义，改译为"methods"。
5. 明确句子的主谓成分。科技文体句式较长，应明确句子真正的主语和谓语。句子"系统采用双伺服驱动……实现对两个……"中的主语应为"system"。所以原译文"… the synchronization control of two servo driver is carried out…"应译为"… the system realizes the synchronization control of two servo driver…"。
6. 保持主谓一致。句子"The test result show that"应译为"The test results show that"；宾语从句"system runs accurately and reliably, and implement the angle"应译为"the system runs accurately and reliably, and implements the angle"。

Unit 8　Agriculture Economy

Text A

Optimistic Future for Agriculture Predicted

Despite these recent uncertainties, "up" is precisely the direction an Iowa State researcher believes agriculture is headed for at least the next 10 years.

Wally Huffman, professor in agricultural economics and Charles F. Curtiss, distinguished professor in Agriculture and Life Sciences, predict supply will go up, demand will go up, and real prices of grain and oilseeds also will go up.

"I'm very optimistic about the next 10 years," said Huffman.

Huffman presented his research to the Organization for Economic Co-operation and Development in Paris, France, last month. OECD and the Iowa Agricultural Experiment Station supported the research.

An important part of Huffman's study was the long-term trend of corn and soybean yields in Iowa, wheat in Kansas and France, rice in Japan and potatoes in the Netherlands. Huffman examined the trends and then made projections about the next decade.

The optimism starts with the producers.

"Prices right now for corn and soybeans are up about 50 percent relative to two years ago, so those are relatively good prices," he said. "That's good news for grain producers."

The impact that the rising demand for biofuels will have on the market for agricultural products is not entirely clear, but grain and oilseed prices will generally be higher than they would be without biofuels.

"Overall, biofuels are probably a good thing for farmers," he said. "However, there will be more erratic variation in grain and oilseed prices than there would be without biofuels," he said.

The main reasons are the erratic components to both supply and demand of crude oil.

While biofuels are pushing demand for grain and oilseeds up, Huffman says, the long-term trend in supply of grain and oilseeds is due to new technologies that are being developed by the private sector and marketed to farmers.

"Supply is going up, and demand is going up," he said. "I think they will grow at a similar pace. There will be occasional spikes due to bad weather and abrupt restriction in crude oil production, but prices will come down. When they do, they will come down to similar levels to what they are now in real terms, and those are pretty good prices."

"For the past 100 years, on average, real agricultural product prices have been falling as technology has been allowing supply to increase faster than demand," he said.

But for the past decade, demand has been rising as quickly as supply, he added.

Yields for major field crops in major producing areas have been steadily increasing. There is no indication that the rate is slowing and no reason to fear falling crop yields. Huffman predicts that the rate of increase in yields for corn and soybeans in major production areas will rise much faster than it has in the past 50 years.

"In the case of corn, since 1955 the average rate of increase in Iowa crop yield has been two bushels, per acre, per year," said Huffman. "That's an amazing accomplishment starting from about 65 bushels, per acre, per year in 1955, up to about 165 bushels, per acre, per year now."

Huffman thinks the future will be even better.

"From 2010 to 2019, corn yields are going to increase quite substantially, maybe at four to six bushels, per acre, per year," he said.

Much of the increase will be due to genetic improvements in hybrid corn varieties associated with new, multiple stacking of genes for insect protection and herbicide tolerance that will permit a major increase in plant population.

These improvements are the result of corn that has been genetically modified (GM) to have certain desirable traits.

Also, better equipment, improved farm management, and reduced- and no-till farming will contribute to rising corn yields in the Midwest.

Other commodities have also improved yield and will likely see continuing increases, according to Huffman.

Soybean yields in Iowa also are increasing, although less dramatically than corn, says Huffman.

The trend over the past 50 years is an increase of about 0.5 bushels, per acre, per year (bu/ac/yr). That rate of improvement in Iowa soybean yields will continue or possibly increase over the next decade. Current soybean yields are about 50 bu/ac/yr.

Kansas is the leading producer of wheat in the United States with yields of about 45 bu/ac/yr. Yields have been improving at about 0.5 bu/ac/yr since about 1950.

Farmers in France are producing wheat at about 113 bu/ac/yr. Yields are improving at more than 1.5 bu/ac/yr.

France is the leading wheat producer in the European Union, and Huffman attributes much of their production advantage to the French emphasis on wheat advantage. They are also showing faster production improvement. France often puts their best land into wheat production. Huffman predicts wheat yields may increase faster if GM wheat is more successful.

Japan is a major rice producer. Yields are improving at a rate of 0.5 bu/ac/yr, and are now at 113 bu/ac/yr compared to around 90 bu/ac/yr in 1960. GM rice has been tried, but has not measurably increased yields, according to Huffman.

The Netherlands is the most advanced country in the world when it comes to potato production technology. Yields in the Netherlands have been increasing by about 4.6 bu/ac/yr over the last 50 years and are now at 670 bu/ac/yr.

"Potatoes are a major world food crop and they don't get a lot of attention," said Huffman. "They are consumed in large amounts in Europe and other places, including the United States, and yields are phenomenal."

Several variables will impact the future of crops.

According to Huffman, the biggest are:

—both private companies and government researchers are working on improving production;

——higher yields as a result of new techniques in breeding crops, including methods to condense—decades of breeding and testing into a few years;

——change in biofuels from corn-based to biomass-based by 2019;

——GM crops gain more acceptance in Europe.

(995 words)

New Words

1. projection [prə'dʒekʃən] *n.* 预测；规划；投影
2. erratic [i'rætik] *adj.* 不稳定的
3. spike [spaik] *n.* 价格（数量）的突然上升；尖状物
4. abrupt [ə'brʌpt] *adj.* 突然的，意外的
5. bushel ['buʃl] *n.* 蒲式耳（谷物、蔬菜容量单位，在美国等于35.2 L）
6. substantially [səb'stænʃəli] *adv.* 大幅度地
7. stacking ['stækiŋ] *n.* 叠加
8. measurably ['meʒərəbli] *adv.* 显著地；可视地
9. phenomenal [fi'nɔminl] *adj.* 非凡的；惊人的
10. condense [kən'dens] *vi.* 浓缩
11. biomass ['baiəumæs] *n.* 生物量

Useful Expressions

1. distinguished professor 知名教授
2. private sector 私营企业
3. in real terms 扣除物价因素；按实质计算
4. field crop 大田作物
5. major producing areas/major production areas 主产区
6. genetic improvement 遗传改良
7. plant population 植物群体
8. desirable trait 优良性状
9. farm management 农田管理
10. no-till farming 免耕法

Notes

1. 本文选自世界贸易组织（World Trade Organization）官网（http://www.WTO.org）。

2. 生物燃料（biofuel），泛指由生物质组成或萃取的固体、液体或气体燃料，可以替代由石油制取的汽油和柴油，是可再生能源开发利用的重要方向。所谓生物质，是指利用大气、水、土地等通过光合作用而产生的各种有机体，即一切有生命的可以生长的有机物质，它包括植物、动物和微生物。不同于石油、煤炭、核能等传统燃料，这种新兴的燃料是可再生燃料。

3. 大田作物（field crop）是指露地种植，直接供给粮食、油料和衣物原料的农作物，主要包括稻谷、小麦、玉米、高粱、油菜、青稞、豆类、棉花等。

Exercises

Part I Vocabulary and Structure

Section A Match each word with its Chinese equivalent.

1. entrepreneur   A. 一体化
2. combination   B. 保护主义
3. integration   C. 有形资产
4. specialization   D. 基石
5. tangible   E. 专业化
6. cornerstone   F. 停滞，萧条
7. protectionism   G. 基本建设
8. infrastructure   H. 合并
9. scarcity   I. 企业家
10. stagnation   J. 短缺

Unit 8 Agriculture Economy

Section B Fill in the blanks with the words or expressions given below. Change the form where necessary.

| genetic | spike | erratic | hybrid | condense |
| measurably | breed | abrupt | phenomenon | project |

1. The recession brought an _____ halt to this happiness.
2. The offspring contain a mixture of the _____ blueprint of each parent.
3. Economists emphasize _____ quantities—the number of jobs, the per capita income.
4. Douglas was a 29-year-old journeyman fighter, _____ in his previous fights.
5. We have learnt how to _____ serious messages into short, self-contained sentences.
6. There is potential for selective _____ for better yields.
7. Africa's mid-1993 population is _____ to more than double by 2025.
8. Exports of Australian wine are growing at a(n) _____ rate.
9. That cell _____ can be used to dissect regulatory mechanisms controlling gene expression in eukaryotic cells（真核细胞）.
10. Although you'd think business would have boomed during the war, there was only a small _____ in interest.

Part Ⅱ Translation

Section A Translate the following sentences into Chinese.

1. The main reasons are the erratic components to both supply and demand of crude oil.
2. For the past 100 years, on average, real agricultural product prices have been falling as technology has been allowing supply to increase faster than demand.
3. These improvements are the result of corn that has been genetically modified (GM) to have certain desirable traits.
4. Soybean yields in Iowa also are increasing, although less dramatically than corn.

5. Netherlands is the most advanced country in the world when it comes to potato production technology.

Section B　Translate the following sentences into English.

1. 他根本没能表现出一个强有力的领导者的样子。
2. 公司的这个计划假定小麦产量会稳定增加。
3. 这一地区现在吸引了60多种鸟类到此繁育后代。
4. 随着空气上升，其温度降低，凝结出水分。
5. 我们难以应对近来出现的需求激增。

Text B

Sustainable Agriculture: Perennial Plants Produce More; Landscape Diversity Creates Habitat for Pest Enemies

Perennial plants（多年生植物）**produce more, require less input than annual croplands.**

The major crops used globally to feed people and livestock—wheat, rice, maize and soy—are based on an annual system, in which crop plants live one year, are harvested, and are replanted the following year. These systems are *notorious*（臭名昭著的）, however, for *stripping*（剥夺）organic nutrients from soils over time.

Perennial systems, on the other hand, contain plants that live longer than one year despite being harvested annually. Many agricultural scientists, including Jerry Glover of The Land Institute, say that perennial crops are the key to creating more sustainable agricultural systems.

"Across agricultural history, we've fundamentally relied on annual grain crops," Glover says. "But at the same time we rely on them, they're degrading the ecosystems they're in, which reduces their productivity."

To compare the long-term sustainability of these two cropping systems, Glover and his colleagues conducted a study on the physical, biological and chemical differences between annual wheat fields and perennial grass fields in Kansas. The fields had each been harvested annually for the past 75 years.

In each test, the researchers found perennial fields to be healthier and more sustainable ecosystems. In the perennial fields, the plants' total root mass was more than seven times that of the annuals, and the roots *infiltrated* (浸润) about a foot deeper into the ground. The perennial fields also had higher soil microbe biodiversity and higher levels of dissolved carbon and nitrogen in the soil. All these findings, says Glover, suggest that the perennial field soil is healthy enough to maintain high levels of organic nutrients.

In addition to being more ecologically sustainable, Glover's team found that the perennial fields were more energy-efficient in providing productive harvests. Although only the annual fields received yearly fertilizer inputs, the perennial fields yielded 23 percent more nitrogen harvested over the 75 years, despite requiring only 8 percent of the energy inputs in the field—such as fertilizer and harvesting operations—as the annual systems.

Glover says that these results clearly show the need to move away from annual crops and increase our use and domestication of perennial crops.

"So far, little effort has been made to improve perennial crops," he says. "Some of greatest possibilities for transforming agriculture may well come from overlooked systems such as perennial grasses."

Landscape diversity creates habitat for pest enemies.

Farmers spend millions of dollars each year on pesticides to kill crop-eating insects. But these insects have natural enemies, too, and new research is investigating what farmers can do to encourage the *proliferation* (增殖) of these pest-eaters. One study, presented as a talk at the ESA meeting, shows that increasing the natural habitat in and around farms can boost populations of pests' natural enemies.

Rebecca Chaplin-Kramer of the University of California Berkeley surveyed the abundance of flies, *ladybugs* (瓢虫), *wasps* (黄蜂) and other *predators* (捕食者) of a common agricultural pest, the cabbage *aphid* (蚜虫), in croplands ranging from 2 percent to about 80 percent natural vegetation. She found that as the proportion of natural area—or complexity—increased, so did the numbers of natural enemies in the croplands.

Chaplin-Kramer shows that increases in predators didn't always result in fewer aphids in the croplands, but she points out that agents of control are only half of the equation and sources of the pests themselves must also be considered. In the absence of predators, pest levels would likely rise even more dramatically.

"By having complexity, you're supplying a community of insects to that farm that will be waiting when—and if—more pests show up," Chaplin-Kramer says.

Fostering larger predator communities is time-consuming and can take years to show results, Chaplin-Kramer says, which is why many farmers are *skeptical* of the idea. But, she says, there's no doubt that a strong predator base is more sustainable than simply using pesticides.

"Pesticides are a short-term solution, because pests can build up resistance, and new pesticides are constantly being developed," she says. "Building up predator communities takes time, but the systems are more stable and will provide more ecosystem services in the long term."

Reduced *tilling* (耕作) improves soil microbe biodiversity.

The idea of using biological instead of chemical methods to create healthy croplands doesn't include just above-ground approaches. Soil bacteria can affect the growth and **success** of crop plants by fixing nitrogen, aiding in the uptake of nutrients and decomposing dead organic matter. Some current farming practices, however, may *disrupt* (破坏) the soil ecosystem and decrease the effectiveness of the microbe community.

In his poster, Shashi Kumar of Texas Tech University will explore the relationship between conventional tilling and low-tilling practices on farms in *semi-arid* (半干旱) areas of west Texas. In areas where soil tilling was kept at a minimum, Kumar and his colleagues found a higher diversity of soil bacteria; conventional tilling produced lower bacterial diversity.

Kumar says that conventional tillage systems disrupt soil particles and decrease soil pore size, which can lead to decreased water and soil access for microbes. Although he recognizes that tillage is necessary, he thinks that farmers can reduce their tillage, even in semi-arid regions, to promote soil bacterial biodiversity.

"We are currently using so many different crop management systems, like pesticides, insecticides and *fungicides* (杀菌剂), which are damaging to our

Unit 8 Agriculture Economy

soil system," Kumar says. "Why shouldn't we focus on biological methods, since the bacteria are already there?"

(937 words)

Comprehension of the Text

Choose the best answer to each of the following questions.

1. Different from the production of major crops, perennial systems are more likely to _____ .
   A. feed people because of the shorter cultivation
   B. create sustainability of agricultural diversity
   C. reduce the fertility of crops
   D. interfere with agricultural production annually
2. Compared with yearly fields, why are the perennial crop fields more fertile?
   A. Because they need little nutrition to gain harvest.
   B. Because they are more likely to create agricultural transformation.
   C. Because they have higher levels of soil organic nutrients.
   D. Because they are characterized with lower-level nitrogen.
3. Which of the following statements is true on the basis of the passage?
   A. Annual maize field has much huger root system than perennial grass field.
   B. Although being more ecologically sustainable, annual fields can't provide more productive yields.
   C. The perennial fields harvest outputs more, requiring less energy inputs.
   D. The annual crops need more nitrogen to harvest because of their high levels of nutrients.
4. The italicized and boldfaced word "*skeptical*" in paragraph 13 most probably means _____ .
   A. ignorant   B. trusting   C. inexperienced   D. incredulous
5. What is the favorable side of the pest-eaters' proliferation?
   A. Farmers are uncertain with the populations of predators.
   B. The environment will gain more benefits from it in the long run.
   C. Predator communities will leave no space for pest subsistence.
   D. Chemical methods, instead of biological ones, will create healthy croplands.

6. Why aren't the pesticides the final solution in the long run?

   A. Because pests can develop resistance to pesticides.

   B. Because new chemical technologies cannot be worked out to produce effective pesticides.

   C. Because experiments on pesticides of crops should be short-term.

   D. Because the perennial fields need not be protected by pesticides.

7. What does the italicized and boldfaced word "success" in paragraph 15 most probably refer to in the text?

   A. Fulfillment.　　B. Productivity.　　C. Performance.　　D. Prosperity.

8. According to the passage, conventional farming practices contribute to _____.

   A. a high level of bacterial percentage

   B. a greater diverse intake of nutrients

   C. unavoidable damages of crop fields

   D. a higher diversity of chemicals

9. Which of the following can influence the number of pest enemies?

   A. Vegetation.　　B. Chemicals.　　C. Habitat.　　D. Fertilizer.

10. It can be inferred from the last two paragraphs that _____.

    A. the conventional tillage systems are necessary in the long term

    B. the bacterial biodiversity can be promoted by pesticides

    C. the chemical crop management approaches are unnecessary in semi-arid regions

    D. the biological crop management methods are the finial solution in the long run

科技文体翻译技巧（八）

标题的译法

一、标题概述

科技文体的标题，应尽量做到以简明扼要的字句揭示文章主题和概括文章内容。此类文章的标题多采用偏正式结构、联合结构、连动结构、动宾结构、陈述句式、疑问句式等结构。翻译时可根据不同情况采取不同的译法，但译得好与坏，对译作的质量有一定影响。只有言简意赅、准确到位的标题翻译，才能帮助读者迅速确定文章内容，成为读者的向导，节约读者时间。

二、科技英语标题种类

（一）并列结构标题

顾名思义，此类标题是由两个或两个以上平行的实词或词组构成，即彼此之间不存在说明和被说明、修饰和被修饰关系的标题。由于此类标题中的各个组成部分的地位是完全平等的，翻译时可按原文的先后次序译成汉语的并列结构标题。

【例】World Energy and Environment Protection

世界能源与环境保护

（二）"前置定语 ＋ 中心语"结构标题

此类标题中的前置定语大多数是名词（名词性词组）、形容词、分词等，这类定语简洁明快，表意完整，能准确说明中心语的意义，相当于汉语的偏正词组。

【例】Agricultural Long-term Test

农业长期试验

（三）"中心语 ＋ 后置定语"结构标题

在科技英语中，此类结构的标题最为多见。其后置短语可以是介词短语、不定式短语、分词短语、定语从句、副词等。翻译时，应将这些后置定语逆序译成中心语的前置词（包括汉语的主谓词组、动宾词组、偏正词组、联合词组等做定语），从而将整个标题译成汉语的偏正结构标题。

1. "中心语 ＋ 介词短语"结构标题

【例】Assessment Technology for Commodities Market

商品市场的评估技术

这种结构的翻译要注意以下几点。

①介词短语后置定语一律译为汉语的"……的"结构，介词本身的意义没有直接显示在字面上。但在许多情况下，介词本身的意义却要译出来。

【例】Investigation on Market Gardening

关于商品蔬菜种植业的研究

②修饰中心词的两个介词短语和中心语之间有固定的搭配关系，译时应按逻辑关系调整原文的语序。

【例】Application to Space Operations of Free-flying Controlled Streams of Liquids

可控自由飞行液流对空间操作的应用

③中心语与其后的介词有固定搭配关系时,介词一般不需译出,而把整个介词短语做中心语的前置定语。

【例】An Approach to Arable Land Cost Estimation

耕地成本估算法

④有时中心语的中心词(一般为名词)前有名词或形容词定语,后有相搭配的介词。此时大多应根据汉语习惯和逻辑关系调整中心词前后定语的语序。

【例】Energy Role in China's Agriculture

能源在中国农业中的作用

⑤根据逻辑意义,中心语的分词或形容词前置定语也偶尔可译成中心语,而把其余部分译作前置修饰语。

【例】Predicted Performance of Shell Structure

壳体结构性能预测

2. "中心语 ＋ 分词短语"结构

【例】An Analysis of Issues Concerning "Acid Rain"

酸雨问题的分析研究

3. "中心语 ＋ 不定式"结构

【例】Practical Ways to Reduce the Cost of Energy

降低能源成本的可行途径

4. "中心语 ＋ 定语从句、副词等"结构

【例】The Acid Smut that Crossed the Atlantic

横穿大西洋的酸性煤尘

(四)"介词短语"结构标题

此类文章标题是"介词 ＋ 中心词"结构。介词大多是 on、about、of、toward 等。这类标题一般可有以下两种译法。

1. 减去介词不译,把介词的宾语(中心语＋后置定语)按以上的方法译成汉语的偏正结构

【例】From Activating Enzyme to Adaptive Enzyme

从活化酶到适应酶的演变

2. 把介词译出,常可译为"关于""试论""谈谈"等,从而把整个标题译作汉语的动宾结构或介词结构

【例】On the Units of the Equilibrium Constant

试论平衡常数单位

（五）"动名词短语"结构标题

1. 很多动名词短语结构的标题

在这种情况下，表达的重心是动名词本身，因此一般动名词译成中心语，而把其宾语部分逆序译成前置修饰语，从而把整个标题译成汉语的偏正结构。如果动名词短语有另一状语修饰，则可译为汉语的连动结构。

【例】Increasing Mileage with a Microprocessor Shift Indictor

采用微处理机换挡指示器增大里程

2. 较短的动名词短语

在这种情况下，可译成汉语的动宾结构（不过这种情况不多见）。

【例】Exploring Larva by Mulberry Leaves

探索桑收法

（六）断开式标题

科技英语中有一类标题不连续，中间有冒号、逗号或破折号隔开，称为断开式标题。这种标题用标点隔开的前后两部分之间存在着各种不同的语义关系，可有以下几种处理方法：

1. 标点前的部分对标点后的部分起限制或说明的作用

在这种情况下，可将标点前部分译成前置修饰语，把标点部分译为中心语，整个标题译成汉语的偏正结构。

【例】Focus：Nuclear Power Plant Outages

核电站停运问题的焦点

2. 标点的前后两部分系同位关系

此即标点后部分对标点前部分进行具体说明，可以保留原标题的结构。

【例】Computers Today and Tomorrow—the Microcomputer Explosion

计算机的现状和未来——计算机的迅猛发展

有时，标点前后两部分非同位关系，而是后部分对前部分进行补充解释、细节补述、背景交代、分层次说明等。这时，为了保持原标题的层次，仍可保留其原有形式不变。译出后实际上相当于汉语的主标题加副标题形式。

【例】Flue-gas Conditioning：Key Advances in Recent Years

烟气处理：近年来的主要改进

（七）陈述句标题

英语中还有一些标题具有一套完整的主谓结构，属陈述句型标题。根据汉语习惯，翻译时可有以下几种处理办法。

1. 标题是"主语＋系动词＋表语"结构

在这种情况下，可把表示性质特征的表语译为主语的前置修饰语，而主语译作中心语，而系动词则减去不译，从而把整个句子标题译为汉语的偏正结构。

【例】The Cataphoretic Chromatography Is Under Development
   研制中的电泳层析谱法

2. 谓语为实义动词的句子标题

在这种情况下，从逻辑上分析谓语动词往往是表达的中心。因此，可把该谓语译成中心语，而把主语和宾语（如谓语动词是及物动词）译成中心语的前置修饰语。从而把整个标题译为汉语的偏正结构。

【例】Computer Program Predicts Condenser Cleanliness
   预测凝汽器清洁度的计算机程序

3. 标题难以变通处理成汉语的偏正结构

在这种情况下，可按原结构译成汉语的陈述句。

【例】Electronic Devices Expend Transmission Flexibility
   电子设备提高了输电的灵活性

（八）疑问句标题

在英语科技报道和科普读物中，常会见到一些疑问句标题。这种标题新颖别致、生动形象，寓知识性和趣味性于一身，很能引起读者的注意力。为了保持这种标题的鲜明特点，翻译时一般可按原形式译为汉语的问句。

【例】Is Machine Intelligence Possible?
   机器会有智能吗？

在不影响原文含义的前提下，某些疑问句标题主要是特殊疑问句，偶可译成汉语的偏正结构标题。这样译时，可把疑问词减去不译，另加一个适当的中心语。一般说来，疑问句标题不宜译为其他形式，因为这会影响原标题的表现力和感染效果，变得枯干苍白，索然无味。

以上讨论的标题译法，原则上也适用于正文中的小标题以及图表标题的翻译处理。

科技英语摘要写作（八）

摘要缩写（Ⅱ）

摘要缩写的主要特点是简明扼要，因此写摘要时要对全文进行选择取舍，抓住重点，突出新观点，强调文章的目的和主要结论。

选择取舍是摘要缩写的关键。主要删减或简化一般描述性、说明性和修饰性语言。缩写后的摘要不仅要简洁，还应和原文一致，保持整体内容的完整流畅。这里介绍第二种摘要缩写方法。

综合概括法

有些英文摘要，基本内容要点分散在摘要之中，条理不清晰，这就需要对分散的要点进行集中概括。现举例如下。

On the Subject and Characteristics of Ideological and Political Education

【原摘要】

Abstract：This paper focuses on the three controversies concerning the subject, the object and their relation in ideological and political education. They are as follows. Is it necessary to introduce this philosophical dichotomy to the discipline of ideological and political education? Under what conditions can the educated become the subject? How should we view the nature of ideological and political education? The author holds that the controversies can be accounted for by the divergence in methodology. Thus, in order to define the subject-object relation, it is imperative to clarify the methodology that has been applied as follows. First, at certain stage of the research, it is necessary to discriminate the parts from the whole; second, when examining the subject-object relation, it is necessary to define the subject and the object from specific perspectives and their mutual relation.

在以上《对思想政治教育主体及其特性的思考》的摘要中，作者把摘要四要素若隐若现地融于整个摘要中，并没有明确说明，且有详有略。要缩写以上英文摘要就要采用综合概括法，重新总结整理。

【缩写后摘要】

Abstract：The purpose of this study is to analyze the subject and characteristics of ideological and political education. The scope of the research covers three controversial methods on the subject and the object. It is concluded that different methodological analyses contribute to the controversies.

摘要译写示例

示例一

【摘要中文原文】

摘要：吉林省是我国重要的绿色农产品产地，绿色农产品产业发展到了一定的规模。为了进一步提高绿色农产品产业的竞争能力，吉林省应建立以维护品牌公信力作为持续发展核心的绿色农产品营销战略体系。为此，应建立绿色农产品质量标准体系、质量认证体系、质量监管体系等组成的绿色农产品营销战略组织体系；完善绿色农产品市场体系和生产体系；采取优品优质的产品策略，采用地理标志与产品标志相结合的品牌策略和差异化定价策略，构建多渠道并进的网络化渠道，利用传统和现代媒介并进的广告宣传策略，将吉林省绿色农产品推向全国高端市场和国际市场。

【摘要原版译文】

Abstract：Jilin Province is the nation's leading producer of green agricultural products. Its Green agricultural products industry has been developed to a certain scale. It is necessary for Jilin Province to set up marketing strategy of green agricultural products, in order to improve the competing capabilities. Thus, Jilin Province should build the specification system, quality certification system, quality supervision and management system. Also, it's better for Jilin Province to adopt superior quality production strategy, Geographical indication strategy and Differential pricing strategy, and to Construct multi-channel network channels, and make use of the advertising strategy through traditional and modern media.

【摘要修改译文】

Abstract：Jilin province is China's leading producer of green agricultural products. The green agricultural product industry in Jilin province has been developed to a certain scale. To improve competitive power, it is necessary for Jilin province to set up marketing strategies of green agricultural products. Thus, Jilin province should build specification system, quality certification system, quality supervision and management system. In addition, it's better for Jilin province to adopt superior quality production strategy, geographical indication strategy and differential pricing strategy, and to construct multi-channel network channels, and make use of the advertising strategy through traditional and modern media.

【主要修改意见】

1. 区分字母的大小写。句中"Green"、"Geographical"和"Construct"不是专有名词也不在句首,首字母应小写。
2. 句子结构合理化。不定式做目的状语,一般置于句首。文中"It is necessary…, in order to improve…"可译为"To improve…, it is necessary…"。
3. 语言表达书面语化。科技文体不用口语表达形式,"Also"应译为"In addition"。

示例二

【摘要中文原文】

　　摘要:本文构建了农业生产综合效益评价指标体系,运用主成分分析法对吉林省农业生产综合效益进行了时间序列分析和横向对比分析,得出的结论是:近年来吉林省的农业生产综合效益不断提高,发展势头良好,但与先进省份相比,仍存在差距。最后从相关方面提出对策建议。

【摘要原版译文】

　　Abstract: Using the method of principal component analysis, the study establishes comprehensive benefit evaluation index system of agricultural production to conduct time series analysis and horizontal comparison analysis. The conclusion is that, the comprehensive benefits of agricultural production in Jilin province continuously improved in recent years, but there are still gaps compared with advanced provinces. Finally, some countermeasures are put forward to improve the comprehensive benefits of agricultural production.

【摘要修改译文】

　　Abstract: Using the method of principal component analysis, we have established comprehensive benefits evaluation index system of agricultural production to conduct time series analysis and horizontal comparison analysis. The conclusion is that the comprehensive benefits of agricultural production in Jilin province have continuously been improved in recent years, but there are still gaps between Jilin province and advanced provinces. Finally, some countermeasures are put forward to improve the comprehensive benefits of agricultural production.

【主要修改意见】

1. 分析逻辑关系。原版译文第一句中"using"的逻辑主语应为人，故应将其逻辑主语改为"we"，时态也应改为完成时态为宜。
2. 恰当使用语态。原版译文第二句中的表语从句的逗号应删除，语态以被动语态为宜，因为动词"improve"的主语为"benefits"。
3. 正确使用时态。原版译文"the comprehensive benefits of agricultural production in Jilin province continuously improved in recent years"运用过去时，不能准确表达"近年来……"的时态，应译为"the comprehensive benefits of agricultural production in Jilin province have continuously been improved in recent years"。

主 要 参 考 文 献

陈世丹,2010. 新编 MPA 英语阅读教程 [M]. 北京:中国人民大学出版社.
范莹芳,2010. 新编科技英语阅读教程 [M]. 哈尔滨:哈尔滨工业大学出版社.
蒋悟生,2010. 生物专业英语 [M]. 北京:高等教育出版社.
李光立,张文芝,2007. 研究生英语阅读教程 [M]. 北京:中国人民大学出版社.
李淑静,金衡山,2007. 专业硕士研究生英语 [M]. 北京:北京大学出版社.
李仪奎,姜名瑛,1992. 中药药理学 [M]. 北京:中国中医药出版社.
李照国,张庆荣,2009. 中医英语 [M]. 上海:上海科学技术出版社.
刘爱军,王斌,2007. 科技英语综合教程 [M]. 北京:外语教学与研究出版社.
秦狄辉,2007. 科技英语写作 [M]. 北京:外语教学与研究出版社.
滕玉梅,胡铁生,2006. 研究生英语读写教程 [M]. 长春:吉林出版集团.
王秉金,1993. 论英语倍数增减的表示方法 [J]. 中国科技翻译(2):1-9.
王慧莉,刘文宇,曹晓玮,2009. MBA 研究生英语综合教程 [M]. 北京:中国人民大学出版社.
王玉雯,吴江梅,2008. 新世纪研究生英语教程 [M]. 北京:北京理工大学出版社.
许修宏,2008. 资源环境科学专业英语 [M]. 北京:中国农业出版社.
许忠能,KEN CHAN,2009. 科技英语 [M]. 北京:清华大学出版社.
郁仲莉,王耀庭,2001. 英语写作与翻译实用教程 [M]. 北京:中国农业出版社.
张兰威,李佳新,李铁晶,等,2007. 食品科学与工程英语 [M]. 哈尔滨:哈尔滨工程大学出版社.
张永萍,吴江梅,2002. 农林英语 [M]. 北京:北京大学出版社.
赵萱,郑仰成,2006. 科技英语翻译 [M]. 北京:外语教学与研究出版社.
郑福裕,徐威,2008. 英文科技论文写作与编辑指南 [M]. 北京:清华大学出版社.
中华人民共和国国家药典委员会,2005. 中国药典 [S]. 北京:化学工业出版社.
周一兵,嵇纬武,2009. 科技英语教程 [M]. 北京:科学出版社.
ANDERSON JAMES G,1992. Health care in the People's Republic of China:A blend of traditional and modern [J]. Central Issues in Anthropology,10:67-75.
HO PENG YOKE,FPETER LIAOWAKI,1997. A brief history of Chinese medicine [M]. Singapore:World Scientific Publishing.
LEVI-STRAUSS CLAUDE,1962. The savage mind [M]. Chicago:University of Chicago Press.
MB MS,S H,2012. Yin and yang:The physical and the symbolic in Chinese medical practices [J]. The University of Western Ontario Journal of Anthropology,20(1).

http://www.econpapers.repec.org

http://www.joe.org/joe/2005june/a6.php

http://www.microbeworld.org/history-of-microbiology

http://www.WTO.org

https://files.eric.ed.gov/fulltext/EJ1308142.pdf

https://opentextbc.ca/introductiontosociology/chapter/chapter1-an-introduction-to-sociology

https://www.agronomy.org

https://www.crops.org/about-crops

https://www.sciencedirect.com/

https://www.scirp.org/journal/paperinformation.aspx?paperid=107598

https://www.uagc.edu/blog/5-principles-of-great-management

https://www.un.org/development/desa/pd/sites/www.un.org.development.desa.pd/files/desa-pd-technicalpaper-living_arrangements_of_older_persons_2019.pdf